高等学校"十三五"规划教材

Advanced Carbon Nanomaterials
（新型碳纳米材料）

Edited by　Zhao Tingkai

（赵廷凯）

西北工业大学出版社

西　安

【Introduction of Content】 Advanced carbon nanomaterials, especially fullerene, carbon nanotubes and graphene are the hottest research topics in recent years. The investigations of fullerene, carbon nanotubes and graphene drive a full develop in the fields of new process, new method and new technology etc. and affect science technology and social development. This textbook consists of six chapters selected from the published literatures and partially from our research achievements. The coverage of this book is as follows: first two chapters are the introduction of nanotechnology and nanomaterials, and carbon and carbon nanomatcrilas. Then the next four chapters deal more with the historical development, synthesis, properties(such as physical and chemical properties) and finally several applications of fullerene, carbon nanotubes and graphene.

This textbook was written to serve as one of the senior undergraduate, graduate or overseas student courses in the fields of new materials, new energy and nanotechnology, and second as a resource and reference book for material scientist and chemists and other researchers working in this field.

图书在版编目(CIP)数据

新型碳纳米材料＝Advanced Carbon Nanomaterials：英文/赵廷凯主编. —西安:西北工业大学出版社,2017.3
 ISBN 978-7-5612-5280-2

Ⅰ.①新… Ⅱ.①赵… Ⅲ.①碳—纳米材料—研究—英文 Ⅳ.①TB383

中国版本图书馆 CIP 数据核字(2017)第 063313 号

策划编辑：杨　军
责任编辑：胡莉巾

出版发行：	西北工业大学出版社
通信地址：	西安市友谊西路 127 号　　邮编:710072
电　　话：	(029)88493844　88491757
网　　址：	www.nwpup.com
印 刷 者：	陕西宝石兰印务有限责任公司
开　　本：	787 mm×1 092 mm　　1/16
印　　张：	8.875
字　　数：	212 千字
版　　次：	2017 年 3 月第 1 版　　2017 年 3 月第 1 次印刷
定　　价：	29.00 元

PREFACE

Nanotechnology is almost a household word nowadays, or at least some words with "nano" in it, such as nano-scale, nano-particle, nano-phase, nano-crystal, or nano-machine. This field now enjoys worldwide attention. Especially, carbon nanomaterials appear to take a lead in their applications.

This field owes its parentage to investigations of reactive species (free atoms, clusters, reactive particles) throughout the 1970s and 1980s, coupled with new techniques and instruments (focus ion beams, innovations in mass spectrometry, vacuum technology, microscopes, and more). Excitement is high and spread throughout different fields, including chemistry, physics, material science, engineering and biology. This is warranted because nanoscale materials represent a new realm of matter, and the possibilities for interesting basic science as well as useful technologies for society are widespread and real.

In spite of all this interest, there is a need for textbooks that serve the basic science community, especially material scientists and chemists.

This textbook "Advanced Carbon Nanomaterials" is written to serve both as one of the textbooks for senior undergraduate, graduate or overseas student courses on carbon-based nanomaterials, and as a resource and reference for material scientists and chemists and other researchers working in this field. Therefore, the readers will find that the chapters are written as a teacher might teach the subject, and not simply as a reference work. For this reason, I hope that this textbook will be adopted for teaching numerous advanced courses in nanotechnology, materials chemistry, and related subjects.

The coverage of this textbook is as follows: first, a detailed introduction of nanotechnology and a brief historical account are given. The next chapters deal more with the synthesis, structure and properties of fullerenes, carbon nanotubes and graphene, such as chemical properties, physical properties, and finally a short chapter on applications of carbon nanomaterials.

The author gratefully acknowledge the contributing authors of these literatures, who are world renowned experts in this burgeoning field of nanotechnology. Their enthusiasm and hard work are very much appreciated. The author also acknowledge the help of my students: Dr. Zhao Xing, Dr. Liu Lehao, Mr. Ji Xianglin et al, as well as my families (especially my wife Liang Jing and my daughter Zhao Xin) for their patience and understanding.

Zhao Tingkai
NPU, Xi'an
2016.12

CONTENTS

Chapter 1 Introduction ··· 1
   1.1 Fundamental concept of nano ··· 1
   1.2 Fundamental concept of nanometer ····································· 1
   1.3 Nanotechnology ··· 3
   1.4 Nanomaterials ·· 11
   Activities and problems for students ·· 17

Chapter 2 Carbon nanomaterials ··· 18
   2.1 Fundamental concepts of carbon ·· 18
   2.2 Allotropes of carbon ··· 22
   2.3 The development of carbon nanomaterials ························· 33
   2.4 The classification of carbon nanomaterials ························· 33
   Activities and problems for students ·· 36

Chapter 3 Fullerenes ·· 38
   3.1 Prediction and discovery of fullerene ································· 38
   3.2 Naming of fullerene ··· 39
   3.3 The discovery of C_{60} ·· 40
   3.4 The synthesis and separation of fullerenes ························· 42
   3.5 The structure confirmation of fullerenes ···························· 44
   3.6 Chemistry of fullerenes ·· 47
   3.7 The doped treatment and application of fullerenes ············ 51
   3.8 Possible applications of fullerenes and their derivatives ···· 53
   Activities and problems for students ·· 56

Chapter 4 Carbon nanotube ·· 57
   4.1 History of carbon nanotube ··· 57
   4.2 The discovery of carbon nanotube ···································· 58

4.3　The type of carbon nanotube ·· 62

4.4　The synthesis of carbon nanotube ·· 62

4.5　The growth mechanism of carbon nanotubes ···································· 73

4.6　Purification methods of carbon nanotube ··· 77

4.7　The micro-characterization and properties of carbon nanotube ········ 79

Activities and problems for students ··· 83

Chapter 5　The applications of carbon nanotubes ·· **84**

5.1　Overview of potential and current applications ································ 84

5.2　Functionalized carbon nanotubes ·· 88

5.3　Applications of carbon nanotubes in energy conversion ················· 90

5.4　Application of carbon nanotubes in military ···································· 92

5.5　Application of carbon nanotubes in electrochemistry ····················· 94

Activities and problems for students ··· 98

Chapter 6　Graphene ·· **99**

6.1　History and discovery of graphene ·· 99

6.2　Mother of all graphitic forms ·· 100

6.3　Structure and properties of graphene ·· 102

6.4　The synthesis methods of graphene ·· 109

6.5　Graphene growth with great size by CVD ······································ 112

6.6　Potential applications of graphene ·· 120

6.7　The future of graphene ·· 122

Activities and problems for students ··· 124

References ··· **125**

Chapter 1 Introduction

1.1 Fundamental concept of nano

Nano is a prefix to a word. It is similar with "kilo" or "macro" etc. Generally, nano as a prefix is used in the unit of length, such as nano-meter (metre). The ten SI prefixes are shown in the chart below (see Table 1-1).

Table 1-1 List of SI prefixes

10^n	Prefix	Symbol	Since	Short scale	Long scale	Decimal
10^{12}	tera	T	1960	Trillion	Billion	1,000,000,000,000
10^9	giga	G	1960	Billion	Milliard	1,000,000,000
10^6	mega	M	1960	Million		1,000,000
10^3	kilo	k	1795	Thousand		1,000
10^1	deca	da	1795	Ten		10
10^0	(none)	(none)	NA	One		1
10^{-3}	milli	m	1795	Thousandth		0.001
10^{-6}	micro	μ	1960	Millionth		0.000,001
10^{-9}	nano	n	1960	Billionth	Milliardth	0.000,000,001
10^{-12}	pico	p	1960	Trillionth	Billionth	0.000,000,000,001

1.2 Fundamental concept of nanometer

Nanometer (abbr. nm) is a unit of length.

From Table 1-1, one nanometer (nm) is one billionth, or 10^{-9}, of a meter. By comparison, typical carbon-carbon bond lengths, or the spacing between these atoms in a molecule, are in the range 0.12 - 0.15 nm, and a DNA double-helix has a diameter around 2 nm. On the other hand, the smallest cellular life-forms, the bacteria of the genus mycoplasma, are around 200 nm in length. They make a comparison with a human hair, which is about 80,000 nm wide (see Fig.1-1).

To put that scale in another context, the comparative size of a nanometer to a meter is the same as that of a marble to the size of the earth. Or another way of putting it: a nanometer is the amount a man's beard grows in the time it takes him to raise the razor to his face.(see Fig.1 - 2 and Fig.1 - 3)

Fig. 1 - 1 Human hair fragment and a network of single-walled carbon nanotubes

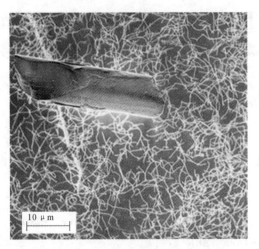

Fig.1 - 2 Size comparisons of nanocrystals with bacteria, viruses and molecules

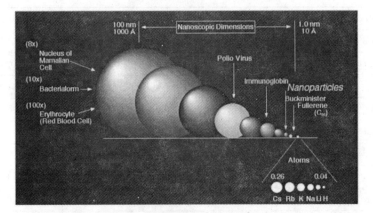

Fig.1 - 3 Size comparisons of nanometer

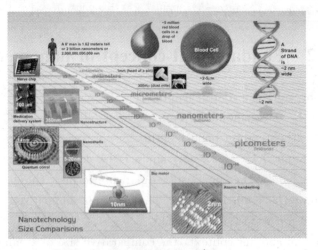

Chapter 1 Introduction

> Why people are interested in the nanoscale?

People are interested in the **nanoscale** (which we define to be **from 100 nm down to the size of atoms (approximately 0.2 nm)**) because it is at this scale that the properties of materials can be very different from those at a larger scale.

Their bulk properties of materials often change dramatically with nano ingredients. Composites made from particles of nano-size ceramics or metals smaller than 100 nanometers can suddenly become much stronger than predicted by existing materials-science models. For example, metals with a so-called grain size of around 10 nanometers are as much as seven times harder and tougher than their ordinary counterparts with grain sizes in the hundreds of nanometers. The causes of these drastic changes stem from the weird world of quantum physics. The bulk properties of any material are merely the average of all the quantum forces affecting all the atoms. As you make things smaller and smaller, you eventually reach a point where the averaging no longer works.

The properties of materials can be different at the nanoscale for **two main reasons**:

(1) Nanomaterials have a relatively larger surface area when compared to the same mass of material produced in a larger form. This can make materials more chemically reactive (in some cases materials that are inert in their larger form are reactive when produced in their nanoscale form), and affect their strength or electrical properties.

(2) Quantum effects can begin to dominate the behaviors of matter at the nanoscale-particularly at the lower end-affecting the optical, electrical and magnetic behaviors of materials. Materials can be produced that are nanoscale in one dimension (for example, very thin surface coatings), in two dimensions (for example, nanowires and nanotubes) or in all three dimensions (for example, nanoparticles).

1.3 Nanotechnology

Neal Lane

"If I were asked for an area of science and engineering that will most likely produce the breakthroughs of tomorrow, I would point to nanoscale science and engineering." Professor Neal Lane, Rice University, former assistant to the President for science and technology and Director of the White House Office of Science and Technology Policy.

The term "nanotechnology" was defined by Tokyo Science University professor Norio Taniguchi in a 1974 paper as follows: "'nano-technology' mainly consists of the processing of separation, consolidation, and deformation of materials by one atom or by one molecule."

1.3.1 What is nanotechnology?

> So what exactly is nanotechnology?

Nanotechnology, shortened to "nanotech", **is the study of the control of matter on an**

atomic and molecular scale. **Generally nanotechnology deals with structures of the size** 100 **nanometers or smaller, and involves developing materials or devices within that size.** Nanotechnology is very diverse, ranging from novel extensions of conventional device physics, to completely new approaches based upon molecular self-assembly, to developing new materials with dimensions on the nanoscale, even to speculation on whether we can directly control matter on the atomic scale.

There has been much debate on the future of implications of nanotechnology. Nanotechnology has the potential to create many new materials and devices with wide-ranging applications, such as in medicine, electronics and energy production. On the other hand, nanotechnology raises many of the same issues as with any introduction of new technology, including concerns about the toxicity and environmental impact of nanomaterials and their potential effects on global economics, as well as speculation about various doomsday scenarios. These concerns have led to a debate among advocacy groups and governments on whether special regulation of nanotechnology is warranted. Another important criteria for the definition is the requirement that the **nanostructure** is man-made. Otherwise you would have to include every naturally formed biomolecule and material particle, in effect redefining much of chemistry and molecular biology as "nanotechnology".

The most important requirement for the nanotechnology definition is that the nanostructure has special properties that are exclusively due to its nanoscale proportions.

1.3.2 What's nanostructure?

A nanostructure is an object of intermediate (middle) size between molecular and microscopic (micrometer-sized) structures. (Note: micro-: on a small scale; macro-: on a large scale)

In describing nanostructures it is necessary to differentiate between the numbers of dimensions on the nanoscale.

(1) **One dimension** on the nanoscale, i.e., only the thickness of the surface of an object is between 0.1 nm and 100 nm.

(2) **Two dimensions** on the nanoscale, i.e., the diameter of the nanotube is between 0.1 nm and 100 nm; its length could be much greater.

(3) **Three dimensions** on the nanoscale, i.e., the spherical nanoparticles is between 0.1 nm and 100 nm in each spatial dimension. The terms nanoparticles and ultrafine particles (UFP) often are used synonymously although UFP can reach into the micrometre range.

Two main approaches are used in nanotechnology, i.e., "bottom-up" and "top-down". In the "bottom-up" approach, materials and devices are built from molecular components which assemble themselves chemically by principles of molecular recognition. In the "top-down" approach, nano-objects are constructed from larger entities without atomic-level control.

Novel areas of physics such as nanoelectronics, nanomechanics and nanophotonics have been evolved during the last decades to provide a basic scientific foundation of nanotechnology.

We found a good definition that is practical and unconstrained by any arbitrary size limitations:

The design, characterization production, and application of structures, devices, and systems by controlled manipulation of size and shape at the nanometer scale (atomic, molecular and macromolecular scale) that produces structures, devices, and systems with at least one novel/superior characteristic or property.

1.3.3 How to characterize in nanoscale?

> What tools & techniques were used?

Nanotechnology and nanoscience got started in the early **1980s** with two major developments: one is the birth of cluster science and the invention of the **scanning tunneling microscope** (STM, see Fig.1-4 and Fig.1-5). (G. Binning (left) & H. Rohrer (right), 1986 Nobel Prize). This development led to the discovery of fullerenes in 1985 and carbon nanotubes a few years later.

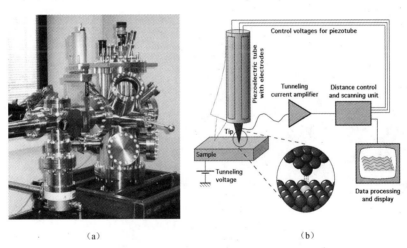

(a)   (b)

Fig. 1-4 Schematic view of STM
(a) optical picture; (b) mechanism diagram

Fig. 1-5 ST Mmage of reconstruction on a clean Au(100) surface

There are other types of scanning probe microscopy, all flowing from the ideas of the scanning confocal microscope developed by Marvin Minsky in 1961 and the scanning acoustic microscope (SAM) developed by Calvin Quate and coworkers in the 1970s, that made it possible to see structures at the nanoscale. The tip of a scanning probe can also be used to manipulate nanostructures (a process called positional assembly). Feature-oriented scanning-positioning methodology suggested by Rostislav Lapshin appears to be a promising way to implement these nanomanipulations in automatic mode. However, this is still a slow process because of low scanning velocity of the microscope. Various techniques of nanolithography such as optical lithography, X-ray lithography, dip pen nanolithography, electron beam lithography or nanoimprint lithography were also developed. Lithography is a top-down fabrication technique where a bulk material is reduced in size to nanoscale pattern.

In another development, the synthesis and properties of semiconductor nanocrystals was studied, which led to a fast increasing number of metal and metal oxide nanoparticles and quantum dots. The atomic force microscope (AFM, see Fig.1-6 and Fig.1-7) was invented six years after the STM was invented. (1986 G. Binning, C.F. Quate & C. Gerber, Stanford University).

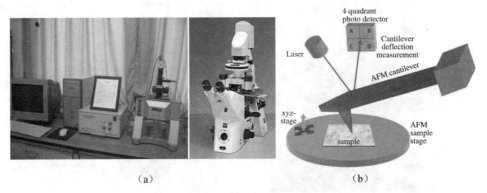

Fig. 1-6 Typical AFM setup
(a)optical pictures; (b)mechanism diagram

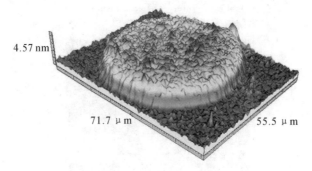

Fig. 1-7 3D AFM image of a DNA biochip

(1) Scanning Electron Microscope (SEM, see Fig. 1-8).
(2) Transmission Electron Microscope (TEM, see Fig. 1-9).

(3) Focused Ion Beam (FIB, see Fig. 1-10 and Fig.1-11).
(4) X-Ray Diffraction (XRD).
(5) Raman Spectroscopy (RS, see Fig. 1-12).
(6) Scanning Probe Microscopes (SPM).
(7) X-ray Photoelectron Spectroscopy (XPS).

(a)　　　　　　　　　　　　(b)

Fig. 1-8　SEM opened sample chamber and SEM micrographs of pollen grains
(a) optical picture;　(b) SEM image of sample

 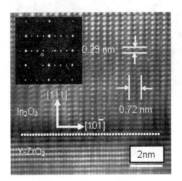

Fig. 1-9　Transmission electron microscope
(a) A basic TEM;　(b) TEM image of Y-ZrO$_2$

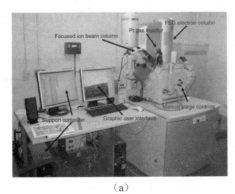 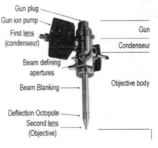

(a)　　　　　　　　　　　　(b)

Fig. 1-10　Focused ion beam
(a) Optical photo of FIB facility;　(b) Schematic diagram of FIB

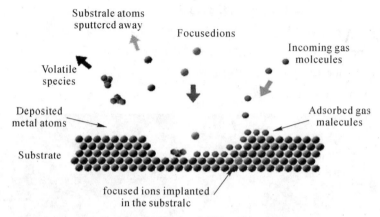

Fig. 1-11 Gas assisted FIB etching process
1. Adsorption of the gas molecules on the substrate.
2. Interaction of the gas molecules with the substrate. Formation of volatile and non volatile species.
3. Evaporation of volatile species and sputtening of non volatile species

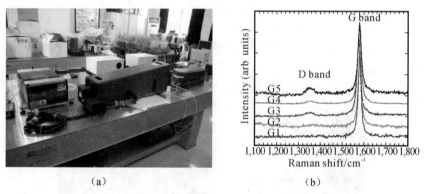

(a)   (b)

Fig. 1-12 Raman spectroscopy
(a) Typical Raman setup; (b) Raman spectrum

1.3.4 History of nanotechnology

Nanotechnology is derived from the Greek words "**nanos**", which means **dwarf**, and **technologia** which means **systematic treatment of an art**. The first use of the concepts in "nanotechnology" (but pre-dating use of that name) was in "There's Plenty of Room at the Bottom — an Invitation to Enter a New Field of Physics" a talk given by physicist Richard Feynman at an American Physical Society meeting at Caltech on December 29, 1959. Feynman described a process by which the ability to manipulate individual atoms and molecules might be developed, using one set of precise tools to build and operate another proportionally smaller set, and so on down to the needed scale. In the course of this, he noted, scaling issues would arise from the changing magnitude of various physical phenomena: gravity would become less important, surface tension and van der

Chapter 1 Introduction

Waals attraction would become more important, etc. This basic idea appears plausible, and exponential assembly enhances it with parallelism to produce a useful quantity of end products. The term "nanotechnology" was defined by Tokyo Science University Professor Norio Taniguchi in a 1974 paper as follows: "'Nanotechnology' mainly consists of the processing of separation, consolidation, and deformation of materials by one atom or by one molecule." In the 1980s the basic idea of this definition was explored in much more depth by Dr. K. Eric Drexler, who promoted the technological significance of nanoscale phenomena and devices through speeches and the books Engines of Creation: "The Coming Era of Nanotechnology (1986)" and nanosystems: molecular machinery, manufacturing, and computation, and so the term acquired its current sense. Engines of Creation: "The Coming Era of Nanotechnology" is considered the first book on the topic of nanotechnology. Nanotechnology and nanoscience got started in the early 1980s with two major developments: the birth of cluster science and the invention of the scanning tunneling microscope (STM). This development led to the discovery of fullerenes in 1985 and carbon nanotubes a few years later. In another development, the synthesis and properties of semiconductor nanocrystals was studied; this led to a fast increasing number of metal and metal oxide nanoparticles and quantum dots. The atomic force microscope (AFM) was invented six years after the STM was invented. In 2000, the United States National Nanotechnology Initiative was founded to coordinate federal nanotechnology research and development.

We define nanoscience as the study of phenomena and manipulation of materials at atomic, molecular and macromolecular scales, where properties differ significantly from those at a larger scale; and nanotechnologies as the design, characterisation, production and application of structures, devices and systems by controlling shape and size at the nanometer scale. In some senses, nanoscience and nanotechnologies are not new. Chemists have been making polymers, which are large molecules made up of nanoscale subunits, for many decades and nanotechnologies have been used to create the tiny features on computer chips for the past 20 years. However, advances in the tools that now allow atoms and molecules to be examined and probed with great precision have enabled the expansion and development of nanoscience and nanotechnologies. Watch an introduction to nanotechnology, starting with Richard Feynman's classic talk in December 1959 "There's Plenty of Room at the Bottom—an Invitation to Enter a New Field of Physics."

> ➢ A word of caution.

Truly revolutionary nanotechnology products, materials and applications, such as nanorobotics, are years in the future (some say only a few years; some say many years). What qualifies as "nanotechnology" today is basic research and development that is happening in laboratories all over the world. "Nanotechnology" products that are on the market today are mostly gradually improved products (using evolutionary nanotechnology) where some form of nanotechnology enabled material (such as carbon nanotubes, nanocomposite structures or nanoparticles of a particular substance) or nanotechnology process (e.g. nanopatterning or quantum dots for medical imaging) is used in the manufacturing process. In their ongoing quest to improve existing products by creating smaller components and better performance materials, all at a lower cost, the number of companies that will manufacture "nanoproducts" (by this definition) will grow very fast and soon make up the majority of all companies across many industries. Evolutionary nanotechnology should therefore be viewed as a process that gradually will affect most companies and industries. (see Fig.1 – 13 and Fig.1 – 14)

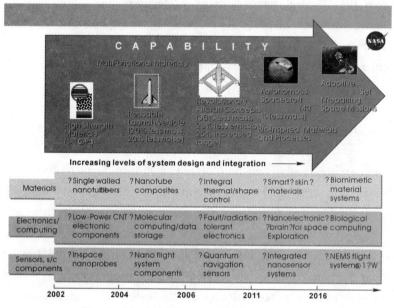

Fig. 1 – 13   NASA nanotechnology roadmap

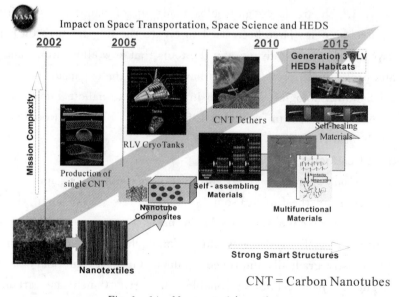

Fig. 1 – 14   Nanomaterials roadmap

1.4 Nanomaterials

An aspect of nanotechnology is the vastly increased ratio of surface area to volume present in many nanoscale materials which makes possible new quantum mechanical effects, for example the "quantum size effect" where the electronic properties of solids are altered with great reductions in particle size. This effect does not come into play by going from macro to micro dimensions. However, it becomes pronounced when the nanometer size range is reached. A certain number of physical properties also alter with the change from macroscopic systems. Novel mechanical properties of nanomaterials are a subject of nanomechanics research. Catalytic activities also reveal new behavior in the interaction with biomaterials.

Nanotechnology can be thought of as extensions of traditional disciplines towards the explicit consideration of these properties. Additionally, traditional disciplines can be re-interpreted as specific applications of nanotechnology. This dynamic reciprocation of ideas and concepts contributes to the modern understanding of this field. Broadly speaking, nanotechnology is the synthesis and application of ideas from science and engineering towards the understanding and production of novel materials and devices. These products generally make copious use of physical properties associated with small scales.

As mentioned above, materials reduced to the nanoscale can suddenly show very different properties compared to what they exhibit on a macroscale, enabling unique applications. For instance, opaque substances become transparent (copper); inert materials attain catalytic properties (platinum); stable materials turn combustible (aluminum); solids turn into liquids at room temperature (gold); insulators become conductors (silicon). Materials such as gold, which is chemically inert at normal scales, can serve as a potent chemical catalyst at nanoscales. Much of the fascination with nanotechnology stems from these unique quantum and surface phenomena that matter exhibits at the nanoscale (gold nanoparticles, AuNPs, see Fig.1 – 15). For example, if you take aluminum and cut it in half, it is still aluminum. But if you keep cutting aluminum in half until it has dimensions on the nano scale, it becomes highly reactive. This is because the molecular structure was changed.

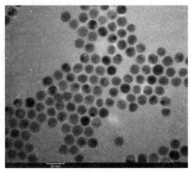

Fig.1 – 15   TEM images of AuNPs

Nanosize powder particles (a few nanometres in diameter, also called nanoparticles) are potentially important in ceramics, powder metallurgy, the achievement of uniform nanoporosity and similar applications. The strong tendency of small particles to form clumps ("agglomerates") is a serious technological problem that impedes such applications. However, a number of dispersants such as ammonium citrate (aqueous) and imidazoline or oleyl alcohol (nonaqueous) are promising solutions as possible additives for deagglomeration.

Nanomaterial is a field which takes amaterials science-based approach to nanotechnology. It studies materials with morphological features on the nanoscale, and especially those which have special properties stemming from their nanoscale dimensions. Nanoscale is usually defined as smaller than a one tenth of a micrometer in at least one dimension, though this term is sometimes also used for materials smaller than one micrometer.

Much of nanoscience and many nanotechnologies are concerned with producing new or enhanced materials. Nanomaterials can be constructed by "top down" techniques, producing very small structures from larger pieces of material, for example by etching to create circuits on the surface of a silicon microchip. They may also be constructed by "bottom up" techniques, atom by atom or molecule by molecule. One way of doing this is self-assembly, in which the atoms or molecules arrange themselves into a structure due to their natural properties. Crystals grown for the semiconductor industry provide an example of self-assembly, as does chemical synthesis of large molecules. A second way is to use tools to move each atom or molecule individually. Although this "positional assembly" offers greater control over construction, it is currently very laborious and not suitable for industrial applications.

It has been 25 years since the scanning tunneling microscope (STM) was invented, followed four years later by the atomic force microscope (AFM), and that's when nanoscience and nanotechnology really started to take off. Various forms of scanning probe microscopes (SPM) based on these discoveries are essential for many areas of today's research. Scanning probe techniques have become the workhorse of nanoscience and nanotechnology research. Here is a scanning electron microscope (SEM) image of a gold tip for near-field scanning optical microscopy (SNOM) obtained by focussed ion beam (FIB) milling. The small tip at the center of the structure measures some tens of nanometers.

Current applications of nanoscale materials include very thin coatings used, for example, in electronics and active surfaces (for example, self-cleaning windows). In most applications the nanoscale components will be fixed or embedded but in some, such as those used in cosmetics and in some pilot environmental remediation applications, free nanoparticles are used. The ability to machine materials to very high precision and accuracy (better than 100 nm) is leading to considerable benefits in a wide range of industrial sectors, for example in the production of components for the information and communication technology, automotive and aerospace industries.

1.4.1 Definition of nanomaterials

Although a broad definition, we categorize nanomaterials as those which have nanostructured components with **at least** one dimension **less than 100 nm.**

(1) Materials that have **one dimension** in the nanoscale (and are extended in the other two dimensions) are layers, such as a thin films or surface coatings. Some of the features on computer chips come in this category.

(2) Materials that are nanoscale in **two dimensions** (and extended in one dimension) include nanowires and nanotubes.

(3) Materials that are nanoscale in **three dimensions** are particles, for example precipitates, colloids and quantum dots (tiny particles of semiconductor materials).

(4) Nanocrystalline materials, made up of **nanometre-sized grains**, also fall into this category. Some of these materials have been available for some time; others are genuinely new.

Two principal factors cause the properties of nanomaterials to differ significantly from other materials: increased relative surface area, and quantum effects. These factors can change or enhance properties such as reactivity, strength and electrical characteristics. As a particle decreases in size, a greater proportion of atoms are found at the surface compared to those inside. For example, a particle of size 30 nm has 5% of its atoms on its surface, at 10 nm 20% of its atoms, and at 3 nm 50% of its atoms. Thus nanoparticles have a much greater surface area per unit mass compared with larger particles. As growth and catalytic chemical reactions occur at surfaces, this means that a given mass of material in nanoparticulate form will be much more reactive than the same mass of material made up of larger particles.

To understand the effect of particle size on surface area, consider a U.S. silver dollar. The silver dollar contains 26.96 g of coin silver, has a diameter of about 40 mm, and has a total surface area of approximately 27.70 cm^2. If the same amount of coin silver were divided into tiny particles — say 1 nm in diameter — the total surface area of those particles would be 11,400 m^2. When the amount of coin silver contained in a silver dollar is rendered into 1 nm particles, the surface area of those particles is 4.115 million times greater than the surface area of the silver dollar!

> What's the difference of nanostructure and nanomaterials?

(1) Nanomaterials are the materials with nanostructure (another definition).

(2) Nanostructure is a structural unit with the size of microstructure between atom and 0.1 μm. (it is a nanomaterial)

(3) "Nanostructured materials" are a subcategory of the term "nanomaterial" (see Fig.1 - 16).

(4) Aggregates and agglomerates are "nanostructured materials", often existing at a micro size, may have some of the behaviour and effects of their smaller subunits, e.g. due to

an increased surface area.

The terms nano-materials, nanostructure in the context of this textbook are meant to cover both "nanomaterials" and "nanostructured materials".

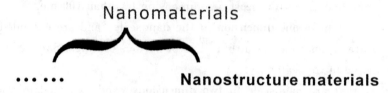

Fig.1 - 16   The terms difference

1.4.2 Size concerns

Another concern is that the volume of an object decreases as the third power of its linear dimensions, but the surface area only decreases as its second power. This somewhat subtle and unavoidable principle has huge ramifications. For example the power of a drill (or any other machine) is proportional to the volume, while the friction of the drill's bearings and gears is proportional to their surface area. For a normal-sized drill, the power of the device is enough to handily overcome any friction. However, scaling its length down by a factor of 1,000, for example, decreases its power by 10,003 (a factor of a billion) while reducing the friction by only 10,002 (a factor of "only" a million). Proportionally it has 1,000 times less power per unit friction than the original drill. If the original friction-to-power ratio was, say, 1%, that implies the smaller drill will have 10 times as much friction as power. The drill is useless.

For this reason, while super-miniature electronic integrated circuits are fully functional, the same technology cannot be used to make working mechanical devices beyond the scales where frictional forces start to exceed the available power. So even though you may see microphotographs of delicately etched silicon gears, such devices are currently little more than curiosities with limited real world applications, for example, in moving mirrors and shutters. Surface tension increases in much the same way, thus magnifying the tendency for very small objects to stick together. This could possibly make any kind of "micro factory" impractical: even if robotic arms and hands could be scaled down, anything they pick up will tend to be impossible to put down. The above being said, molecular evolution has resulted in working cilia, flagella, muscle fibers and rotary motors in aqueous environments, all on the nanoscale. These machines exploit the increased frictional forces found at the micro or nanoscale. Unlike a paddle or a propeller which depends on normal frictional forces (the frictional forces perpendicular to the surface) to achieve propulsion, cilia develop motion from the exaggerated drag or laminar forces (frictional forces parallel to the surface) present at micro and nano dimensions. To build meaningful "machines" at the nanoscale, the relevant forces need to be considered. We are faced with the development and design of intrinsically pertinent machines rather than the simple reproductions of macroscopic ones.

All scaling issues therefore need to be assessed thoroughly when evaluating nanotechnology for practical applications.

1.4.3 Applications

As of August 21, 2008, the Project on Emerging Nanotechnologies estimates over 800 manufacturer-identified nanotech products are publicly available with new ones hitting the market at a pace of 3 - 4 per week. The project lists all of the products in a publicly accessible online inventory. Most applications are limited to the use of "first generation" passive nanomaterials which includes titanium dioxide in sunscreen, cosmetics and some food products; Carbon allotropes used to produce gecko tape, silver in food packaging, clothing, disinfectants and household appliances, zinc oxide in sunscreens and cosmetics, surface coatings, paints and outdoor furniture varnishes; and cerium oxide as a fuel catalyst.

The National Science Foundation (a major distributor for nanotechnology research in the United States) funded researcher David Berube to study the field of nanotechnology. His findings are published in the monograph Nano-Hype: "The Truth Behind the Nanotechnology Buzz". This published study (with a foreword by Mikhail Roco, Senior Advisor for Nanotechnology at the National Science Foundation) concludes that much of what is sold as "nanotechnology" is in fact a recasting of straightforward materials science, which is leading to a "nanotech industry built solely on selling nanotubes, nanowires, and the like" which will "end up with a few suppliers selling low margin products in huge volumes". Further applications which require actual manipulation or arrangement of nanoscale components await further research. Though technologies branded with the term "nano" are sometimes little related to and fall far short of the most ambitious and transformative technological goals of the sort in molecular manufacturing proposals, the term still connotes such ideas. According to Berube, there may be a danger that a "nano bubble" will form, or is forming already, from the use of the term by scientists and entrepreneurs to garner funding, regardless of interest in the transformative possibilities of more ambitious and far-sighted work.

Nano-membranes have been produced that are portable and easily-cleaned systems that purify, detoxify and desalinate water meaning that third-world countries could get clean water, solving many water related health issues.

1.4.4 Implications

Due to the far-ranging claims that have been made about potential applications of nanotechnology, a number of serious concerns have been raised about what effects these will have on our society if realized, and what action if any is appropriate to mitigate these risks.

There are possible dangers that arise with the development of nanotechnology. The Center for Responsible Nanotechnology suggests that new developments could result, among other things in untraceable weapons of mass destruction, networked cameras for use by the government and weapons developments fast enough to destabilize arms races ("Nanotechnology Basics").

One area of concern is the effect that industrial-scale manufacturing and use of nanomaterials would have on human health and the environment, as suggested by nanotoxicology research. Groups such as the Center for Responsible Nanotechnology have advocated that nanotechnology should be specially regulated by governments for these reasons. Others counter that overregulation would stifle scientific research and the development of innovations which could greatly benefit mankind.

Other experts, including director of the Woodrow Wilson Center's Project on Emerging Nanotechnologies David Rejeski, have testified that successful commercialization depends on adequate oversight, risk research strategy, and public engagement. Berkeley, California is currently the only city in the United States to regulate nanotechnology, Cambridge, Massachusetts in 2008 considered enacting a similar law, but ultimately rejected this.

1.4.5 Safety of manufactured nanomaterials

Nanomaterials behave differently than other similarly-sized particles. It is therefore necessary to develop specialized approaches to testing and monitoring their effects on human health and on the environment. The OECD Chemicals Committee has established the Working Party on Manufactured Nanomaterials to address this issue and to study the practices of OECD member countries in regards to nanomaterial safety.

While nanomaterials and nanotechnologies are expected to yield numerous health and health care advances, such as more targeted methods of delivering drugs, new cancer therapies, and methods of early detection of diseases, they also may have unwanted effects. Increased rate of absorption is the main concern associated with manufactured nanoparticles.

When materials are made into nanoparticles, their surface area to volume ratio increased. The greater specific surface area (surface area per unit weight) may lead to increased rate of absorption through the skin, lungs, or digestive tract and may cause unwanted effects to the lungs as well as other organs. However, the particles must be absorbed in sufficient quantities in order to pose health risks.

As the use of nanomaterials increases worldwide, concerns for worker and user safety are mounting. To address such concerns, the Swedish Karolinska Institute conducted a study in which various nanoparticles were introduced to human lung epithelial cells. The results, released in 2008, showed that iron oxide nanoparticles caused little DNA damage and were non-toxic. Zinc oxide nanoparticles were slightly worse. Titanium dioxide caused only DNA damage. Carbon nanotubes caused DNA damage at low levels. Copper oxide was found to be the worst offender, and was the only nanomaterial identified by the researchers as a clear health risk.

In October 2008, the Department of Toxic Substances Control (DTSC), within the

California Environmental Protection Agency, announced its intent to request information regarding analytical test methods, fate and transport in the environment and other relevant information from manufacturers of carbon nanotubes. The term "manufacturers" include persons and businesses that produce nanotubes in California or import carbon nanotubes into California for sale. The purpose of this information request will be to identify information gaps and to develop information about carbon nanotubes, an important emerging nanomaterial.

On January 22, 2009, a formal information request letter was sent to manufacturers who produce or import carbon nanotubes in California, or who may export carbon nanotubes into the State. This letter constitutes the first formal implementation of the authorities placed into statute by AB 289 (2006) and is directed to manufacturers of carbon nanotubes, both industry and academia within the State, and to manufacturers outside California who export carbon nanotubes to California. This request for information must be met by the manufacturers within one year.

1.4.6 What then are the nanomaterials of today and tomorrow?

As a framework for this discussion, the approach of Jones for organizing nanotechnology into three categories can be applied to nanomaterials. The categories are:

(1) Incremental nanotechnology — improving the properties of materials by controlling their nanoscale structure.

(2) Evolutionary nanotechnology — taking a step beyond redesigning simple materials at the nanoscale and designing nanoscale devices that do something interesting.

(3) Radical nanotechnology — developing nanoscale machines that would exist at the convergence of nanotechnology, biotechnology, information technology and cognitive technology.

Activities and problems for students

Activities.

(1) What's the **nano**?
(2) What's nanometer?
(3) What's nanoscale?
(4) What's nanostructure?

Problems.

(1) What's the difference of nanomaterials and nanostructure?
(2) Why do we study nanomaterials?
(3) What's nanotechnology?

Chapter 2   Carbon nanomaterials

2.1   Fundamental concepts of carbon

Carbon (pronounced/karbən/) is the chemical element with symbol C and atomic number 6. As a member of group Ⅳ on the periodic table, it is nonmetallic and tetravalent — making four electrons available to form covalent chemical bonds. There are three naturally occurring isotopes, with ^{12}C and ^{13}C being stable, while ^{14}C is radioactive, decaying with a half-life of about 5,730 years. Carbon is one of the few elements known since antiquity. The name "carbon" comes from Latin language "carbo", coal and in some Romance and Slavic languages, the word carbon can refer both to the element and to coal.

Carbon is one of the abundant elements on the earth. Almost all organics are composed of carbon networks, and carbon materials are very familiar in our daily lives, for example, ink for newspapers, "lead" for pencils, active carbon in refrigerators, and so on. Carbon materials, which consist mainly of carbon atoms, have been used since the prehistoric era as charcoal.

Carbon is one of the least abundant elements in the Earth's crust, but the fourth most abundant element in the universe by mass after hydrogen, helium, and oxygen. It is present in all known life-forms, and in the human body carbon is the second most abundant element by mass (about 18.5%) after oxygen. This abundance, together with the unique diversity of organic compounds and their unusual polymer-forming ability at the temperatures commonly encountered on earth, make this element the chemical basis of all known life (see Fig. 2-1).

Carbon was discovered in prehistory and was known in the forms of soot and charcoal to the earliest human civilizations. Diamonds were known probably as early as 2,500 BC in China, while carbon in the form of charcoal was made around Roman times by the same chemical process as it is today, by heating wood in a pyramid covered with clay to exclude air.

1. What are the differences of Carbon(碳) and Charcoal(炭) ?

Carbon is a nonmetallic element. Charcoal is a substance or compound containing carbon.

There are several allotropes of carbon of which the best known are graphite, diamond and amorphous carbon. The physical properties of carbon vary widely with the allotropic form. For example, diamond is highly transparent, while graphite is opaque and black. Diamond is among the hardest materials known, while graphite is soft enough to form a streak on paper (hence its name, from the Greek word "to write"). Diamond has a very low electrical conductivity, while graphite is a very good conductor. Under normal conditions, diamond has the highest thermal conductivity of all known materials. All the allotropic forms are solids under normal conditions but graphite is the most thermodynamically stable.

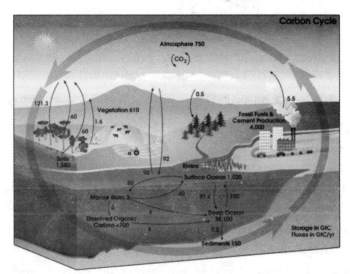

Fig.2-1 Diagram of the carbon cycle

The black numbers indicate how much carbon is stored in various reservoirs, in billions of tons ("GtC" stands for gigatons of carbon; figures are circa 2004). The purple numbers indicate how much carbon moves between reservoirs each year. The sediments, as defined in this diagram, do not include the 70 million GtC of carbonate rock and kerogen (image from NASA)

All forms of carbon are highly stable, requiring high temperature to react even with oxygen (see Fig.2-2). The most common oxidation state of carbon in inorganic compounds is $+4$, while $+2$ is found in carbon monoxide and other transition metal carbonyl complexes. The largest sources of inorganic carbon are limestones, dolomites and carbon dioxide, but significant quantities occur in organic deposits of coal, peat, oil and methane clathrates. Carbon forms more compounds than any other element, with almost ten million pure organic compounds described to date, which in turn are a tiny fraction of such compounds that are theoretically possible under standard conditions.

A new allotrope of carbon, fullerene, that was discovered in 1985 includes nanostructured forms such as buckyballs and nanotubes. Their discoverers (Curl, Kroto and Smalley) received the Nobel Prize in Chemistry in 1996. The resulting renewed interest in new forms lead to the discovery of further exotic allotropes, including glassy carbon, and the realization that "amorphous carbon" is not strictly amorphous.

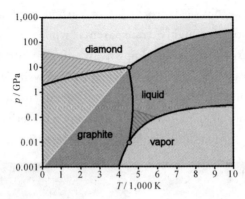

Fig.2-2 Theoretically predicted phase diagram of carbon (image from English Wikipedia)

2. What are the difference of isotope(同位素), isotropy(各向同性) and allotrope(同素异形体)?

Isotope: same atom, different numbers; isotropy: is uniformity in all orientations; allotrope: same molecular, different structure.

As to 2009, graphene appears to be the strongest material ever tested. However, the process of separating it from graphite will require some technological development before it is economical enough to be used in industrial processes.

Atomic carbon is a very short-lived species and therefore, carbon is stabilized in various multi-atomic structures with different molecular configurations called allotropes. The three relatively well-known allotropes of carbon are amorphous carbon, graphite and diamond. Once considered exotic, fullerenes are nowadays commonly synthesized and used in research; they include buckyballs, carbon nanotubes, carbon nanobuds and nanofibers. Several other exotic allotropes have also been discovered, such as lonsdaleite, glassy carbon, carbon nanofoam and linear acetylenic carbon. (see Table 2-1, Fig.2-3 to Fig.2-5)

Table 2-1 The system of carbon allotropes spans a range of extremes

Synthetic nanocrystalline diamond is the hardest materials known	Graphite is one of the softest materials known
Diamond is the ultimate abrasive	Graphite is a very good lubricant
Diamond is an excellent electrical insulator	Graphite is aconductor of electricity
Diamond is the best known naturally occurring thermal conductor	Some forms of graphite are used for thermal insulation (i.e. firebreaks and heat shields)
Diamond is highly transparent	Graphite is opaque
Diamond crystallizes in the cubic system	Graphite crystallizes in the hexagonal system
Amorphous carbon is completely isotropic	Carbon nanotubes are among the most an isotropic materials ever produced

Chapter 2 Carbon nanomaterials

Fig. 2 - 3 Graphite ore

Fig. 2 - 4 Raw diamond crystal

(Images from English Wikipedia)

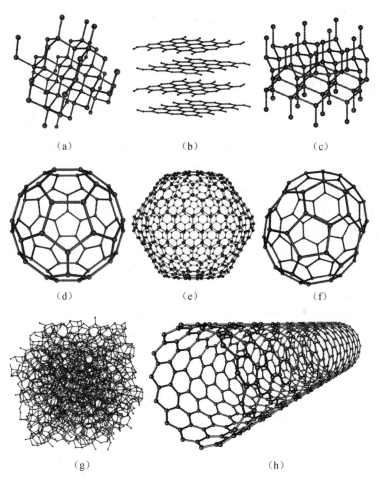

Fig.2 - 5 Structure schematics of the allotropes of carbon
(a) diamond; (b) graphite; (c) lonsdaleite;
(d)-(f) fullerenes (C_{60}, C_{540}, C_{70}); (g) amorphous carbon; (h) carbon nanotube

2.2 Allotropes of carbon

2.2.1 Diamond

Diamond is one of the best known allotropes of carbon, whose hardness and high dispersion of light make it useful for industrial applications and jewelry. Diamond is the hardest known natural mineral, which makes it an excellent abrasive and makes it hold polish and luster extremely well. No known naturally occurring substance can scratch, let alone cut, a diamond.

The market for industrial-grade diamonds operates much differently from its gem-grade counterpart. Industrial diamonds are valued mostly for their hardness and heat conductivity, making many of the gemological characteristics of diamond, including clarity and color, mostly irrelevant. This helps explain why 80% of mined diamonds (equal to about 100 million carats or 20,000 kg annually), unsuitable for use as gemstones and known as bort, are destined for industrial use. In addition to mined diamonds, synthetic diamonds found industrial applications almost immediately after their invention in the 1,950s, another 400 million carats (80,000 kg) of synthetic diamonds are produced annually for industrial use — nearly four times the mass of natural diamonds mined over the same period.

The dominant industrial use of diamond is in cutting, drilling (drill bits), grinding (diamond edged cutters) and polishing. Most uses of diamonds in these technologies do not require large diamonds, in fact, most gem-quality diamonds can find an industrial use. Diamonds are embedded in drill tips or saw blades, or ground into a powder for use in grinding and polishing applications. Specialized applications include use in laboratories as containment for high pressure experiments, high-performance bearings and limited use in specialized windows.

With continuing advances being made in the production of synthetic diamond, future applications are beginning to become feasible. Garnering much excitement is the possible use of diamond as a semiconductor suitable to build microchips from, or the use of diamond as a heat sink in electronics. Significant research efforts in Japan, Europe, and the United States are under way to capitalize on the potential offered by diamond's unique material properties, combined with increased quality and quantity of supply starting to become available from synthetic diamond manufacturers.

Each carbon atom in a diamond is covalently bonded to four other carbons in a tetrahedron. These tetrahedrons together form a 3-dimensional network of 6-membered carbon rings (similar tocyclohexane), in the chair conformation, allowing for zero bond angle strain. This stable network of covalent bonds and hexagonal rings is the reason that diamond is so incredibly strong (see Fig. 2-6).

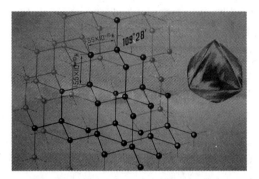

Fig. 2-6 Atom structural drawing of diamond

2.2.2 Graphite

Graphite (named by Abraham Gottlob Werner in 1789, from the Greek γράφειν (graphein, "to draw/write", for its use in pencils), is one of the most common allotropes of carbon. The structure was shown in Fig. 2-7. Unlike diamond, graphite is an electrical conductor, and can be used as the material in the electrodes of an electrical arc lamp. Graphite holds the distinction of being the most stable form of carbon under standard conditions. Therefore, it is used in thermochemistry as the standard state for defining the heat of formation of carbon compounds.

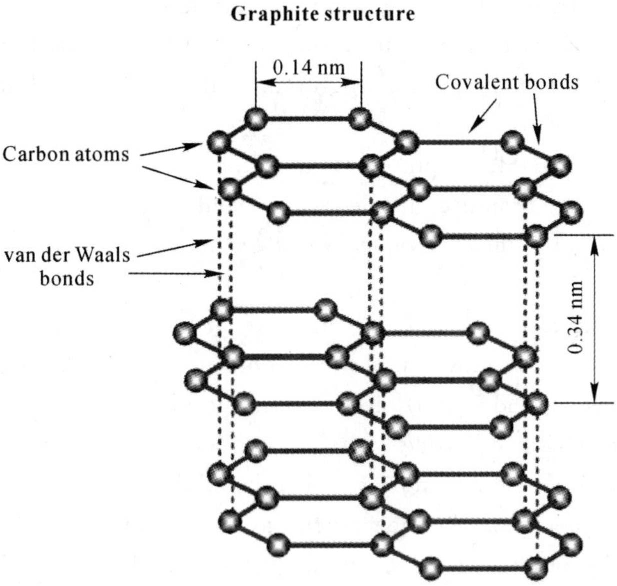

Fig. 2-7 Atom structural drawing of graphite

Graphite is able to conduct electricity, due to delocalization of the pi-bond electrons above and below the planes of the carbon atoms. These electrons are free to move, so are able to conduct electricity. However, the electricity is only conducted along the plane of the layers. In diamond all four outer electrons of each carbon atom are "localised" between the atoms in covalent bonding. The movement of electrons is restricted and diamond does not conduct an electric current. In graphite, each carbon atom uses only 3 of its 4 outer energy level electrons in covalently bonding to three other carbon atoms in a plane. Each carbon atom contributes one electron to a delocalised system of electrons that is also a part of the chemical bonding. The delocalised electrons are free to move throughout the plane. For this reason, graphite conducts electricity along the planes of carbon atoms, but does not conduct in a direction normal to the plane.

Graphite powder is used as a drylubricant. Although it might be thought that this industrially important property is due entirely to the loose interlamellar coupling between sheets in the structure, in fact in a vacuum environment (such as in technologies for use in space), graphite was found to be a very poor lubricant. This fact led to the discovery that graphite's lubricity is due to adsorbed air and water between the layers, unlike other layered dry lubricants such as molybdenum disulfide. Recent studies suggest that an effect called superlubricity can also account for this effect.

When a large number of crystallographic defects bind these planes together, graphite loses its lubrication properties and becomes what is known aspyrolytic carbon, a useful material in blood-contacting implants such as prosthetic heart valves.

Natural and crystalline graphites are not often used in pure form as structural materials due to their shear-planes, brittleness and inconsistent mechanical properties.

In pure glassy (isotropic) synthetic forms, pyrolytic graphite and carbon fiber graphite are extremely strong, heat-resistant (to 3,000 ℃) materials, used in reentry shields for missile nose-cones, solid rocket engines, high temperature reactors, brake shoes and electric motor brushes.

Intumescent or expandable graphiteis used in fire seals, fitted around the perimeter of a fire door. During a fire the graphite intumesces (expands and chars) to resist fire penetration and prevent the spread of fumes. A typical start expansion temperature (SET) is between 150 ℃ and 300 ℃.

Density: its specific gravity is 2.3, which makes it lighter than diamond.

Effect of heat: it is the most stable allotrope of carbon. At high temperatures and pressures (roughly 2,000 ℃ and 5 GPa), it can be transformed into diamond. At about 700 ℃, it burns in oxygen, forming carbon dioxide.

Chemical activity: it is slightly more reactive than diamond. This is because the reactants are able to penetrate between the hexagonal layers of carbon atoms in graphite. It is unaffected by ordinary solvents, dilute acids, or fused alkalis. However, chromic acid oxidizes it to carbon dioxide.

2.2.3 Lonsdaleite

The diamond is a kind of octahedron crystal which is composed of quadruple linkage carbon atoms. It is known that the diamond is the hardest material which exists for the present. However, the magazine "New Scientist" (USA) reported recently according to

Britain that the diamond is not the hardest material, as Lonsdaleite ("the blue silk black eyebrow coloring stone" in Chinese) is theoretically predicted to be 58% stronger than diamond. Lonsdaleite also has the carbon atom composition, but it is one kind of hexahedron crystal (see Fig. 2-8). The Lonsdaleite color ranges from light brown to yellow, which is extremely rare in the nature. According to the scientist studies, in nature Lonsdaleite sometimes formed when meteorites containing graphite hit the earth. The great heat and stress of the impact transforms the graphite into diamond, but retains graphite's hexagonal crystal lattice and constitutes the cubic hexagonal pattern. In nature the purity of the existing Lonsdaleite is not high, therefore its degree of hardness is generally lower than the diamond, but in the laboratory we can make the Lonsdaleite's degree of hardness 58% higher than the diamond. According to the scientific research test in University of Nevada (USA), Lonsdaleite is simulated to be 58% harder than diamond on the (100) face and to resist indentation pressures of 152 GPa, whereas diamond would break at 97 GPa. Although diamond is rare, when compared to Lonsdaleite, diamond might be everywhere. Therefore, in the foreseeable future, there is no possibility to use the Lonsdaleite in substitution for diamond in a large scale. De Beers Corporation does not need to worry. Over the last century, De Beers has been highly successful in increasing consumer demand for diamonds. The statue of diamond as "girls' best friend" will not vacillate for a long time.

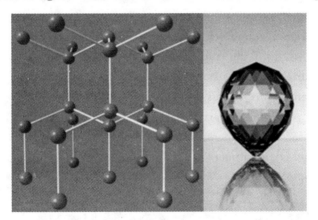

Fig.2-8 Crystal structure of Lonsdaleite

2.2.4 Buckminsterfullerenes

The buckminsterfullerenes, or usually just fullerenes for short, were discovered in 1985 by a team of scientists from Rice University (USA) and the University of Sussex (UK), three of whom were awarded the 1996 Nobel Prize in Chemistry. They are named for the resemblance of their allotropic structure to the geodesic structures devised by the scientist and architect Richard Buckminster "Bucky" Fuller. Fullerenes are molecules of varying sizes and composed entirely of carbon, which take the form of a hollow sphere, ellipsoid or tube (see Fig.2-9).

As of the early 21st century, the chemical and physical properties of fullerenes are still under heavy study, in both pure and applied research laboratories. In April 2003, fullerenes were under study for potential medicinal use — binding specific antibiotics to the structure to target resistant bacteria and even target certain cancer cells such as melanoma.

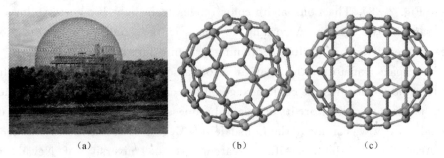

Fig.2 - 9   Geodesic dome structures
(a)image from the website: http://www.planetarium-jena.de/Geschichte.43.0.html; (b)C_{70} model; (c)C_{60}

Since the discovery of fullerenes, structural variations on fullerenes have evolved well beyond the individual clusters themselves. Examples include:

(1) **buckyball clusters**: smallest member is C_{20} (unsaturated version of dodecahedrane) and the most common is C_{60}.

(2) **nanotubes**: hollow tubes of very small dimensions, having single or multiple walls; potential applications in electronics industry.

(3) **megatubes**: larger in diameter than nanotubes and prepared with walls of different thickness; potentially used for the transport of a variety of molecules of different sizes.

(4) **polymers**: chain, two-dimensional and three-dimensional polymers are formed under high pressure high temperature conditions.

(5) **nano "onions"**: spherical particles based on multiple carbon layers surrounding a buckyball core, proposed for lubricants (see Fig.2 - 10).

(6) **linked "ball-and-chain" dimers**: two buckyballs linked by a carbon chain.

(7) **fullerene rings**.

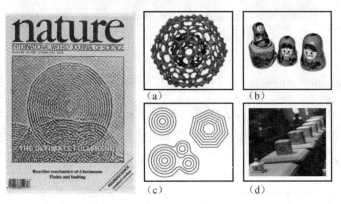

Fig. 2 - 10   Carbon onions (image from the cover of *Nature* 1992, Copyright [1992] *Nature*)
(a)consist of fullerenes contained one inside another, (b)thus resembling in a way the structural principle of a Matryoshka. (c) Apart from perfectly spherical onions, however, there are also faceted species and such with several onions connected in one particle and (d) China Famensi Relic Box.

2.2.5 Amorphous carbon

Amorphous carbon is the name used for carbon that does not have any crystalline structure. As with all glassy materials, some short-distance order can be observed, but there is no long-distance pattern of atomic positions. While completely amorphous carbon can be produced, most amorphous carbon actually contains microscopic crystals of graphite-like or even diamond-like carbon.

Coal, soot and carbon black are informally called amorphous carbon. However, they are products of pyrolysis (the process of decomposing a substance by the action of heat), which does not produce true amorphous carbon under normal conditions. The coal industry divides coal up into various grades, depending on the amount of carbon present in the sample compared to the amount of impurities. The highest grade, anthracite, is about 90% carbon and 10% other elements. Bituminous coal is about 75%–90% carbon, and lignite is the name for coal that is around 55% carbon.

A snapshot of the 64 atom C network was shown in Fig. 2-11. The heavy lines show the network of bonds; the 22 dark spheres depict threefold coordinated atoms (sp^2 hybridized) and the 42 light spheres show the fourfold coordinated atoms (sp^3 hybridized).

The simulations were performed by N. Marks, (Dept of Applied Physics, University of Sydney, in Australia) at the Max Planck Institute, Stuttgart (Germany).

Fig. 2-11 Structural drawing of amorphous carbon

2.2.6 Carbon nanotubes

Carbon nanotubes (CNTs), also called buckytubes, are cylindrical carbon molecules with novel properties that make them potentially useful in a wide variety of applications (e.g., nano-electronics, optics, materials applications, etc.). They exhibit extraordinary strength, unique electrical properties and are efficient conductors of heat. Inorganic nanotubes have also been synthesized. A nanotube is a member of the fullerene structural family, which also includes buckyballs. Whereas buckyballs are spherical in shape, a nanotube is cylindrical, with at least one end typically capped with a hemisphere of the buckyball structure. Their name is derived from their size, since the diameter of a nanotube is on the order of a few nanometers (approximately 50,000 times smaller than the width of a human hair), while they can be up to several centimeters in length. There are two main types of nanotubes: multi-walled carbon nanotubes (MWCNTs) (see Fig. 2 - 12) and single-walled carbon nanotubes (SWCNTs) (see Fig. 2 - 13).

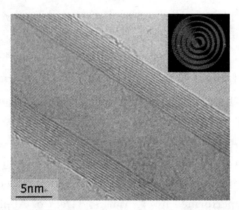

Fig. 2 - 12　HRTEM image of multi-walled carbon nanotubes
(from the website: http://endomoribu.shinshu-u.ac.jp/research/cnt)

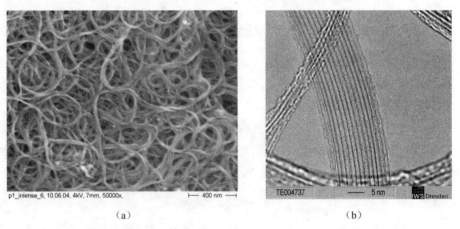

Fig. 2 - 13　SEM and HRTEM images of single-walled carbon nanotubes
(images from Fraunhofer IWS Dresden)
(a)SEM;　(b)HRTEM

2.2.7　Glassy carbon

Glassy carbon or vitreous carbon is a class of non-graphitizing carbon (see Fig. 2 - 14) which is widely used as an electrode material in electrochemistry, as well as for high temperature crucibles and as a component of some prosthetic devices. It was produced by workers at the laboratories of the General Electric Company, UK, in the early 1960s using cellulose as the starting material. A short time later, Japanese workers produced a similar material from phenolic resin.

It was first produced by Bernard Redfern in the mid 1950's at the laboratory of the Carborundum Company, Manchester, UK. He set out to develop a polymer matrix to mirror a diamond structure and discovered a resole (phenolic) resin that would, with special preparation, set without a catalyst. Using this resin the first glassy carbon was produced. Patents were filed some of which were withdrawn in the interests of national security. Original research samples of resin and product exist.

The preparation of glassy carbon (see Fig. 2-15) involves subjecting the organic precursors to a series of heat treatments at temperatures up to 3,000 ℃. Unlike many non-graphitizing carbons, they are impermeable to gases and are chemically extremely inert, especially those which have been prepared at very high temperatures. It has been demonstrated that the rates of oxidation of certain glassy carbons in oxygen, carbon dioxide or water vapor are lower than those of any other carbon. They are also highly resistant to attack by acids. Thus, while normal graphite is reduced to a powder by a mixture of concentrated sulfuric and nitric acids at room temperature, glassy carbon is unaffected by such treatment, even after several months.

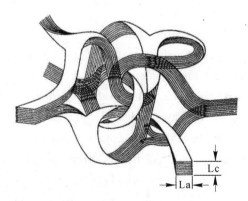

Fig. 2-14 Structure of glassy carbon
(from G.M. Jenkins and K. Kawamura.
1971, Copyright [1971] *Nature*)

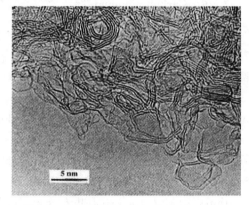

Fig. 2-15 TEM image of glassy carbon
(from the website: http://www.
htw-germany.com/technology)

2.2.8 Carbon nanofoam

Carbon nanofoam is the fifth known allotrope of carbon discovered in 1997 by Andrei V. Rode and co-workers at the Australian National University in Canberra. It consists of a low-density cluster-assembly of carbon atoms strung together in a loose three-dimensional web (see Fig.2-16).

Each cluster is about 6nm wide and consists of about 4,000 carbon atoms linked in graphite-like sheets that are given negative curvature by the inclusion of heptagons among the regular hexagonal pattern. This is the opposite of what happens in the case of buckminsterfullerenes, in which carbon sheets are given positive curvature by the inclusion of pentagons.

The large-scale structure of carbon nanofoam is similar to that of anaerogel, but with 1% of the density of previously produced carbon aerogels only a few times the density of air at sea level. Unlike carbon aerogels, carbon nanofoam is a poor electrical conductor.

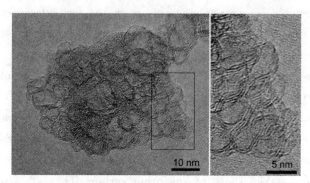

Fig. 2-16 HRTEM images of carbon nanofoam
(from Hideo Kohno et al. 2012. Copyright [2012] *Journal of Nanoscience and Nanotechnoglogy*)

2.2.9 Carbon nanobud

In nanotechnology, carbon nanobuds discovered and synthesized in 2006 (see Fig. 2-17), form a material which combines two previously discovered allotropes of carbon: carbon nanotubes and fullerenes. In this new material fullerenes are covalently bonded to the outer sidewalls of the underlying nanotubes. Consequently, these nanobuds exhibit properties of both carbon nanotubes and fullerenes. For instance, the mechanical properties and the electrical conductivity of these nanobuds are similar to those of corresponding carbon nanotubes. However, because of the higher reactivity of the attached fullerene molecules, the hybrid material can be further functionalized through known fullerene chemistry. Additionally, the attached fullerene molecules can be used as molecular anchors to prevent slipping of the nanotubes in various composite materials, thus modifying the composite's mechanical properties.

Due to the large number of highly curved fullerene surfaces acting as electron emission sites on conductive carbon nanotubes, nanobuds possess advantageous field electron emission characteristics. Randomly oriented nanobuds have already been demonstrated to have an extremely low work function for field electron emission. Reported test measurements show (macroscopic) field thresholds of about 0.65 V·μm^{-1} (non-functionalized SWCNTs have a (macroscopic) field threshold for field electron emission as high as 2 V·μm^{-1}) and a much higher current density as compared with that of corresponding pure SWCNTs.

Properties such as chemical reactivity, good dispersion and variable band gap electronic structure suggest wide applicability of nanobuds. As the production processes are scalable, the nanobud applications may have industrial importance.

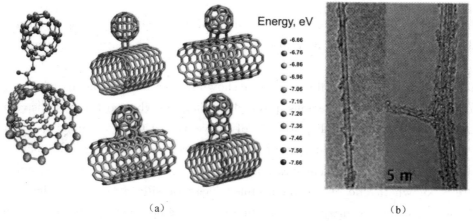

Fig. 2-17 Computer models of several stable nanobud structures and HRTEM of carbon nanobuds (from Albert G et al. 2007. Copright [2007] *Nature Nanotechnology*)
(a) Computer models; (b) HRTEM

2.2.10 Graphene

Graphene is a one-atom-thick planar sheet of sp^2-bonded carbon atoms that are densely packed in a honeycomb crystal lattice. It can be viewed as an atomic-scale chicken wire made of carbon atoms and their bonds. The name comes from "graphite" and "alkene", graphite itself consists of many graphene sheets stacked together (see Fig. 2-18 and details in Chapter 6).

The carbon-carbon bond length in graphene is approximately 0.142nm. Graphene is the basic structural element of some carbon allotropes including graphite, carbon nanotubes and fullerenes. It can also be considered as an infinitely large aromatic molecule, the limiting case of the family of flat polycyclic aromatic hydrocarbons called graphenes.

Measurements have shown that graphene has a breaking strength 200 times greater than steel, making it the strongest material ever tested.

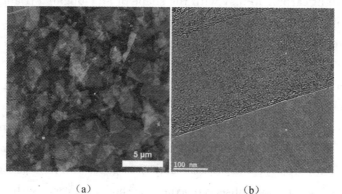

Fig. 2-18 AFM and TEM images of graphene
(images from the website: http://sticky.kaist.ac.kr/menu2/menu5.php)
(a) AFM; (b) TEM

2.2.11 Other possible carbon forms

Chaoite or white carbon: chaoite is a mineral believed to have been formed under meteorite impacts. It has been described as slightly harder than graphite with a reflection color of grey to white. However, the existence of carbyne phases is disputed.

Metallic carbon: Theoretical studies have shown that carbon (diamond) when brought at enormous pressure, there are regions in the phase diagram where it has metallic character.

Cubic carbon: At ultrahigh pressures of above 1,000 GPa, diamond is predicted to transform into the so-called C_8 structure, a body-centered cubic structure with 8 atoms in the unit cell (see Fig. 2-19). This cubic carbon phase might have importance in astrophysics. Its structure is known in one of the metastable phases of silicon and is similar to cubane. Superdense and superhard material resembling this phase has been synthesized recently.

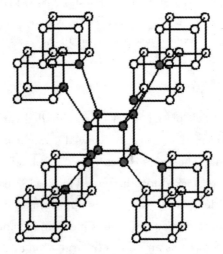

Fig. 2-19 Cubic carbon nanocrystal

Prismane C_8 is a theoretically-predicted metastable carbon allotrope comprising an atomic cluster of eight carbon atoms, with the shape of an elongated triangular bipyramid — a six-atom triangular prism with two more atoms above and below its bases.

Amorphous Diamond: A high-pressure superhard carbon allotrope. Glassy carbon is an amorphous carbon allotrope containing nearly 100% sp^2 bonding at ambient conditions. It has a fullerene-related structure, where fragments of curved graphene like sheets of linked hexagons with dispersed pentagons and heptagons randomly distribute throughout the network. Glassy carbon combines desirable properties of glasses and ceramics with those of graphite, such as high temperature stability, extreme resistance to chemical attack, high proportion of isolated porosity, and impermeability to gases and liquids. Compressing glassy carbon above 40 GPa, we have observed a new carbon allotrope with a fully sp^3-bonded amorphous structure and diamond-like strength. Synchrotron X-ray and Raman spectroscopy revealed a continuous pressure-induced sp^2-to-sp^3 bonding change, while X-ray diffraction confirmed the perseverance of non-crystallinity. The transition was reversible upon releasing pressure. Used as an indentor, the glassy carbon ball demonstrated its exceptional strength by reaching 130 GPa with a confining pressure of 60 GPa. Such an extremely large stress difference of >70 GPa has never been observed in any material except diamond, indicating the high hardness of this high-pressure carbon allotrope.

2.3　The development of carbon nanomaterials

Development and characterization of nanophase materials is the new world-wide research thrusts in advanced materials. Among various nanomaterials developed in the last decade, carbon nanomaterials (nanofibers, tubes and graphene) have attained maximum attraction of researchers as well as industries. This is mainly because of the extraordinary properties of carbon nanomaterials resembling metal, semiconductor as well as superconductor and also for their hydrogen storage capacity. Carbon nanotubes have been synthesized by numerous techniques, such as arc discharge, laser ablation, catalytic CVD etc. With the attractive properties of carbon nanotubes and the potential applications in mind, there have been concerted efforts world-wide to produce these materials in more economical way. In some references, researchers have launched programs to synthesize carbon nanomaterials through physical and chemical methods. Aligned carbon nanomaterials have been synthesized through CVD method by cracking of hydrocarbon gases (benzene, xylene, etc.) in presence of catalyst (Fe, Ni, etc.) at temperature of 700 – 800 ℃. The substrate was prepared through sol-gel route in order to control the size and distribution of catalyst particles. Benzene leads to development of more carbon nanofibers are observed with xylene as precursor. The relative amounts of aligned and randomly oriented carbon nanotubes can be controlled through optimized processing conditions. Nanotubes have also been synthesized using solids (chars and cokes). The precursors are impregnated with a suitable catalyst solution. On heating these precursors to a temperature of 650 – 900 ℃, carbon nanotubes have been observed embedded in amorphous carbons. These carbon nanomaterials have been characterized using XRD, TEM, SEM and thermal analysis techniques.

2.4　The classification of carbon nanomaterials

Since today's device engineering is facing technical and economic difficulty in further miniaturizing electronic devices with the current fabrication technologies, the need for alternative device-channels is particularly imminent. Fullerenes (see Fig.2 – 20) and carbon nanotubes have been considered as the most promising nanomaterials in today's nanoscience and nanotechnology since Kraetschmer and Huffman's first report on the macroscopic synthesis of C_{60} in 1990. In fact, fullerenes and carbon nanotubes have been successfully used for nanometer-sized devices such as diodes, transistors, and random memory cells during the past several years.

Carbon nanomaterials mainly include:

(1) **fullerenes.**

(2) **graphene.**

(3) **carbon nanotubes.**

(4) **carbon nanofibers.**

(5) **exo-and endo-hedral metallofullerenes.**

(6) **carbon nanohorn.**

(7) **carbon nanocage.**

(8) **carbon nanoonions.**

(9) **carbon nanocone.**

(10) **carbon nanopeapods.**

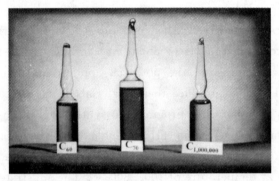

Fig. 2-20 Fullerenes solution with different carbon atoms
(image from the website: http://www.chm.bris.ac.uk/webprojects2001/tweedale/)

Carbon nanohorn: Nanoscale ruthenium/gold bimetallic clusters. It has been used as a molecular precursor to produce pure ruthenium nanoparticles (seeds) as catalysts for the growth of carbon nanohorns (CNHs, see Fig. 2-21).

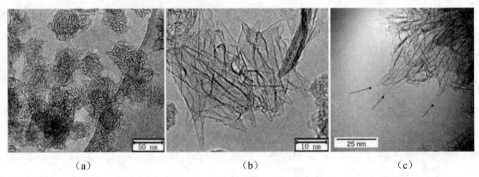

Fig. 2-21 TEM of carbon nanohorn (from N Li et al. 2010. Copyright [2010] *Carbon*)
(a) Low magnificafion TEM; (b) High magnification TEM; (c) Horn tep

Carbon nanocage: It was found that carbon nanocages (CNCs) (see Fig. 2-22) showed excellent performance as direct methanol fuel cell (DMFC) electrodes. CNCs have not attracted much attention until very recent years. Several fabrication routes of CNCs have been developed, such as solid-phase synthesis, self-assembly template approach, poly (methyl methacrylate)/poly(divinylbenzene) (PMMA/ PDVB) or polypyrrole core/shell template synthesis, and surface-modified colloidal silica template process. All these syntheses are complicated multiple-step processes and some of them used poisonous chemical precursors, which prevents them from further scale-up production. Recently, carbon-coated iron nanoparticles with diameters ranging from 5 to 50 nm and a few layers of graphitic shells have been synthesized in a mass scale by laser-induction complex heating evaporation.

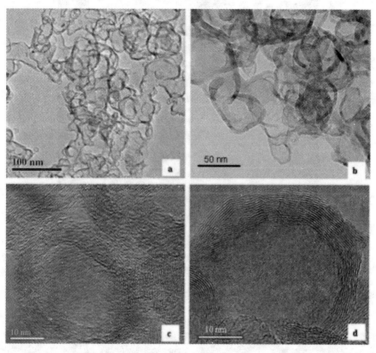

Fig. 2 - 22 HRTEM of carbon nanocages
(from Wang et al. 2007. Copyright [2007] *Journal of Materials Chemistry*)

Carbon nanocones: Carbon nanocones are conical structures that are made predominantly from carbon and have at least one dimension of the order one micron or smaller. Nanocones have height and base diameter of the same order of magnitude, this distinguishes them from tipped nanowires which are much longer than their diameter. Nanocones occur on the surface of natural graphite. Hollow carbon nanocones can also be produced by decomposing hydrocarbons with a plasma torch (see Fig. 2 - 23). Electron microscopy reveals that the opening angle (apex) of the cones is not arbitrary, but has preferred values of approximately 20°, 40° and 60°. This observation was explained by a model of the cone wall composed of wrapped graphene sheets, where the geometrical requirement for seamless connection naturally accounted for the semi-discrete character and the absolute values of the cone angle.

Carbon nanopeapods: Carbon nanotubes encapsulating fullerenes, metallofullerenes and other novel molecules (the so-called "peapods", see Fig. 2 - 24) have been synthesized in high-yield. Such peapods materials have been found to possess not only unique structural properties but more importantly novel electronic transport properties as revealed by high-resolution transmission electron microscopy/electron energy loss spectroscopy (HRTEM-EELS), STM/STS and FET (field effect transistors) measurements. We have found that nano-peapods encaging metallofullerenes exhibit bandgap modulation due to the electron transfer from metallo-fullerenes to carbon nanotubes. Such nano-peapods have been applied to FET with novel device properties.

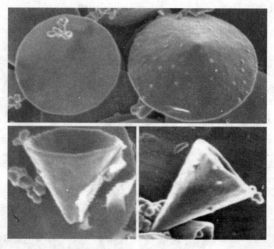

Fig. 2 - 23 SEM images of a carbon disk (top left image) and free-standing hollow carbon nanocones produced by pyrolysis of heavy oil in the Kvaerner Carbon Black & Hydrogen Process. Maximum diameter is about 1 μm
(from Naess S N et al. 2009. Copyright [2009] *Science and Technology of Advanced Materials*)

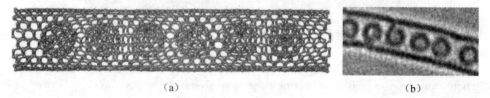

(a) (b)

Fig. 2 - 24 Carbon peapods
(a) A model structure of metal containing fullerenes inserted in single-walled carbon nanotube;
(b) HRTEM image of carbon peapods
(images from the website: http://nano.chem.nagoya-u.ac.jp/research/topics.html)

Activities and problems for students

Activities.

(1) What's carbon?

(2) What's chacoal?

(3) What's allotropes?

Problems.

(1) What's the difference of carbon and charcoal?

(2) What's the difference of isotope, isotropy and allotrope?

(3) What's the type of carbon nanomaterials?

(4) What are the differences of graphite and diamond?

Chapter 3　Fullerenes

3.1　Prediction and discovery of fullerene

The existence of C_{60} was predicted by Eiji Osawa of Toyohashi University of Technology in a Japanese magazine in 1970. He noticed that the structure of a corannulene molecule was a subset of a soccer-ball shape, and he made the hypothesis that a full ball shape could also exist(see Fig. 3-1). His idea was reported in Japanese magazines, but did not reach Europe or America.

With mass spectrometry, discrete peaks were observed corresponding to molecules with the exact mass of sixty or seventy or more carbon atoms. In 1985, Harold Kroto (then of the University of Sussex), James R. Heath, Sean O'Brien, Robert Curl and Richard Smalley (see Fig. 3-2), from Rice University, discovered C_{60}, and shortly thereafter came to discover the fullerenes. Kroto, Curl, and Smalley were awarded the 1996 Nobel Prize in Chemistry for their roles in the discovery of this class of compounds. C_{60} and other fullerenes were later

Fig. 3-1　Japanese bamboo vase which incorporates pentagonal and hetagonal rings Courtesy Prof. Eiji Osawa

noticed occurring outside the laboratory (e.g., in normal candle soot). By 1991, it was relatively easy to produce gram-sized samples of fullerene powder using the techniques of Donald Huffman and Wolfgang Krätschmer. Fullerene purification remains a challenge to chemists and to a large extent determines fullerene prices. So-called endohedral fullerenes have ions or small molecules incorporated inside the cage atoms. Fullerene is an unusual reactant in many organic reactions such as the Bingel reaction discovered in 1993. The first nanotubes were obtained in 1991.

Minute quantities of the fullerenes, in the form of C_{60}, C_{70}, C_{76} and C_{84} molecules are produced in nature hidden in soot and formed by lightning discharges in the atmosphere.

Recently, fullerenes were found in a family of minerals known as Shungites in Karelia, Russia.

Fig. 3-2   Smalley (fifth from left) was the only academic present when President Bush signed the 21st Century Nanotechnology R & D Act in 2003

3.2   Naming of fullerene

Buckminsterfullerene (C_{60}) was named after Richard Buckminster Fuller, a noted architectural modeler who popularized the geodesic dome (see Fig. 3-3). Since buckminsterfullerenes have a similar shape to that sort of dome, the name was thought to be appropriate. As the discovery of the fullerene family came after buckminsterfullerene, the shortened name "fullerene" was used to refer to the family of fullerenes.

For illustrations of geodesic dome structures, see Montreal Biosphère, Eden Project, Missouri Botanical Garden, Science World at TELUS World of Science, Mitchell Park Horticultural Conservatory, Gold Dome, Tacoma Dome, Reunion Tower, and Spaceship Earth (Epcot).

Fig. 3-3   The scientist Richard Buckminsterfuller design the American international reading extensively hall spherical dome shell construction

3.3 The discovery of C_{60}

In the mid-1980s a new class of carbon material was discovered called carbon 60 (C_{60}). Richard Smalley (left), Harry Kroto (middle), and Robert Curl (right) the experimental chemists who discovered C_{60} named it "buckminsterfullerene", in recognition of the architect Buckminster Fuller, who was well-known for building geodesic domes, and the term fullerenes was then given to any closed carbon cage. C_{60} are spherical molecules about 1 nm in diameter, comprising 60 carbon atoms arranged as 20 hexagons and 12 pentagons: the configuration of a football. In 1990, a technique to produce larger quantities of C_{60} was developed by resistively heating graphite rods in a helium atmosphere. Several applications are envisaged for fullerenes, such as miniature "ball bearings" to lubricate surfaces, drug delivery vehicles and in electronic circuits.

Since the discovery of fullerenes in 1985, structural variations on fullerenes have evolved well beyond the individual clusters themselves. Examples include:

(1) **buckyball clusters**: smallest member is C_{20} (unsaturated version of dodecahedrane) and the most common is C_{60}.

(2) **nanotubes**: hollow tubes of very small dimensions, having single or multiple walls; potential applications in electronics industry.

(3) **megatubes**: larger in diameter than nanotubes and prepared with walls of different thickness; potentially used for the transport of a variety of molecules of different sizes.

(4) **polymers**: chain, two-dimensional and three-dimensional polymers are formed under high pressure high temperature conditions.

(5) **nano "onions"**: spherical particles based on multiple carbon layers surrounding a buckyball core, proposed for lubricants (see Fig.3-4).

(6) **linked "ball-and-chain" dimers**: two buckyballs linked by a carbon chain.

(7) **fullerene rings**.

Buckminsterfullerene (IUPAC name (C_{60}) fullerene) is the smallest fullerene molecule in which no two pentagons share an edge (which can be destabilizing, as in pentalene). It is also the most common in terms of natural occurrence, as it can often be found in soot (see Fig. 3-5).

Chapter 3 Fullerenes

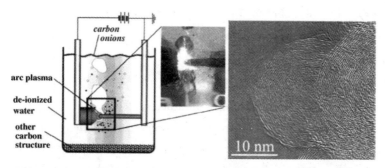

Fig. 3-4 Scheme of the apparatus used for the arc discharge synthesis of CNOs

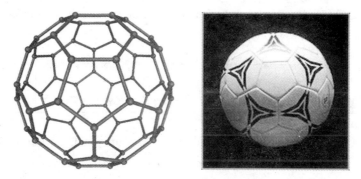

Fig. 3-5 An association football is a model of the Buckminsterfullerene C_{60}

The structure of C_{60} is a truncated ($T=3$) icosahedron, which resembles a soccer ball of the type made of twenty hexagons and twelve pentagons, with a carbon atom at the vertices of each polygon and a bond along each polygon edge.

The van der Waals diameter of a C_{60} molecule is about 1 nm. The nucleus to nucleus diameter of a C_{60} molecule is about 0.7 nm.

The C_{60} molecule has two bond lengths. The 6:6 ring bonds (between two hexagons) can be considered "double bonds" and are shorter than the 6:5 bonds (between a hexagon and a pentagon). Its average bond length is 1.4 angstroms($\text{Å}, 1\text{Å}=10^{-10}$ m).

Silicon buckyballs have been created around metal ions.

1. Boron buckyball

A new type of buckyball utilizingboron atoms instead of the usual carbon has been predicted and described by researchers at Rice University. The B-80 structure, with each atom forming 5 or 6 bonds, is predicted to be more stable than the C-60 buckyball. One reason for this given by the researchers is that the B-80 is actually more like the original geodesic dome structure popularized by Buckminster Fuller which utilizes triangles rather than hexagons. However, this work has been subject to much criticism by quantum chemists as it was concluded that the predicted Ih symmetric structure was vibrationally unstable and the resulting cage undergoes a spontaneous symmetry break yielding a puckered cage with rare Th symmetry (symmetry of volleyball). The number of six atom rings in this molecule is 20 and number of five member rings is 12. There is an additional atom in the center of each six member ring, bonded to each atom surrounding it.

2. Variations of buckyballs

Another fairly common buckminsterfullerene is C_{70}, but fullerenes with 72, 76, 84 and even up to 100 carbon atoms are commonly obtained.

In mathematical terms, the structure of a fullerene is a trivalent convex polyhedron with pentagonal and hexagonal faces. In graph theory, the term fullerene refers to any 3-regular, planar graph with all faces of size 5 or 6 (including the external face). It follows from Euler's polyhedron formula, $|V|-|E|+|F|=2$, (where $|V|$, $|E|$, $|F|$ indicate the number of vertices, edges, and faces), that there are exactly 12 pentagons in a fullerene and $|V|/2-10$ hexagons.

20-fullerene (dodecahedral graph)	26-fullerene graph	60-fullerene (truncated icosahedral graph)	70-fullerene graph

The smallest fullerene is the dodecahedron — the unique C_{20}. There are no fullerenes with 22 vertices. The number of fullerenes C_{2n} grows with increasing $n = 12, 13, 14, \ldots$, roughly in proportion to n^9 (sequence A007894 in OEIS). For instance, there are 1,812 non-isomorphic fullerenes C_{60}. Note that only one form of C_{60}, the buckminsterfullerene alias truncated icosahedron, has no pair of adjacent pentagons (the smallest such fullerene). To further illustrate the growth, there are 214,127,713 non-isomorphic fullerenes C_{200}, 15,655,672 of which have no adjacent pentagons.

Trimetasphere carbon nanomaterials were discovered by researchers at Virginia Tech and licensed exclusively to Luna Innovations. This class of novel molecules comprises 80 carbon atoms (C_{80}) forming a sphere which encloses a complex of three metal atoms and one nitrogen atom. These fullerenes encapsulate metals which puts them in the subset referred to as metallofullerenes. Trimetaspheres have the potential for use in diagnostics (as safe imaging agents), therapeutics and in organic solar cells.

3.4　The synthesis and separation of fullerenes

3.4.1　The apparatus

While the truncated icosahedron structure was compelling and appealing, it could not be confirmed until macroscopic quantities of C_{60} were obtained. The breakthrough came in 1990 when Wolfgang Krchmer (Max Planck Institute, Heidelberg, Germany) and Donald Huffman (University of Arizona, Tucson, AZ) discovered C_{60} in graphitic carbon "soot", produced by evaporating graphite electrodes via resistive heating in an atmosphere of 100

Torr helium (1 atm = 760 Torr)[①](Krchmer and Lamb, 1990).

Although the soot contained only a few percent by weight of C_{60}, it could be conveniently extracted using benzene as solvent. The red-brown benzene solution could be decanted from the black insoluble soot and then dried using gentle heat, leaving a residue of dark brown to black crystalline material. Mass spectral analysis of this material showed peaks at 720 (C_{60}) and 840 (C_{70}) in an approximate ratio of 10 : 1.

Shortly after the Krchmer synthesis was reported, Smalley's group at Rice published a modified design for the "C_{60} generator" (Haufler, 1990). In the Smalley apparatus (see Fig. 3 – 6), an electric arc is maintained between two nearly contacting graphite electrodes. Hence, most of the power is dissipated in the arc and not in resistive heating of the rod. The entire electrode assembly is enclosed in a reaction kettle that is filled with 100 Torr pressure of helium. Black soot, like that observed by Krchmer, is produced, and extraction with organic solvents yields fullerenes.

There are many other methods to prepare fullerenes, such asresistive heating, partial combustion of hydrocarbons, pyrolysis of hydrocarbons and so on.

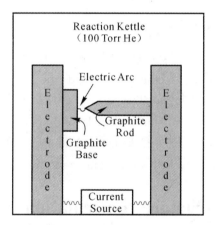

Fig. 3 – 6 Schematic diagram of the contact-arc apparatus used to generate macroscopic quantities of C_{60}

3.4.2 The experimental results

Interestingly, the original goal of the Krchmer/Huffman study was not to generate C_{60}. Rather, it was to generate graphite "soot" in the lab and compare the spectral properties of this material with those of interstellar carbon dust. They initially detected the presence of C_{60} in their soot samples by observing three broad features in the ultraviolet spectrum. In addition, they saw four sharp infrared bands superimposed on a rather large continuum background absorption from the graphitic carbon which comprised $>95\%$ of the sample. (Krchmer & Fostiroupoulos, 1990) Theorists had previously shown that C_{60}, because of its

---

①   1 atm=101.325 kPa,1 Torr=133.322 Pa.

unusually high symmetry, would have only four infrared-active vibrational modes and the positions of the peaks observed by Krchmer et al. closely matched the calculated line positions.

C_{60} is only man-made? (see Fig. 3-7)

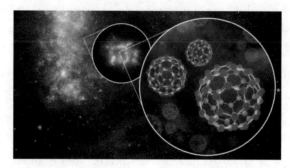

Fig. 3-7 An infrared photo of the Small Magellanic Cloud taken by Spitzer is shown here in this artist's illustration, with two callouts. The middle callout shows a magnified view of an example of a planetary nebula, and the right callout shows an even further magnified depiction of buckyballs, which consist of 60 carbon atoms arranged like soccer balls. Image credit: NASA/JPL-Caltech

3.4.3 Separation of C_{60} and C_{70}

Kroto's group at Sussex was the first to show that the mixture of C_{60} and C_{70}, obtained using the Krchmer synthetic technique, could be cleanly separated by column chromatography (Taylor R., 1990). The Kroto chromatographic separation (alumina, hexane) produced pure C_{60} as a mustard-colored solid that appeared brown or black with increasing film thickness. It gave beautiful magenta solutions. C_{70} was a reddish-brown solid and thicker films were grayish-black. Its solutions were orange or amber.

3.5 The structure confirmation of fullerenes

There are many methods to confirm C_{60} fullerenes, such as:

(1) Nuclear Magnetic Resonance (NMR) Spectra.
(2) Mass spectrometry.
(3) Single Crystal X-Ray Structure Determinations.
(4) Raman spectrometry.
(5) FT-IR.
(6) STM etc.

3.5.1 Nuclear magnetic resonance (NMR) spectra

The four-line IR spectrum for C_{60}, as reported by Krchmer et al. (Krchmer & Lamb, 1990), supported the proposed truncated icosahedron structure. However, even more convincing was the C NMR spectrum of the purified C_{60}, reported by Kroto et al. (Taylor,

1990). The NMR spectrum contained a single peak at 142.7, as expected for the highly symmetrical truncated icosahedron structure in which all carbons are identical. This result eliminated planar graphite fragments and fullerenes of lower symmetry as possible structures for C_{60}. A sixty-membered polyalkyne ring would also be expected to exhibit one ^{13}C NMR signal but the observed chemical shift position (142.7) was inconsistent with this possibility.

The ^{13}C NMR spectrum of purified C_{70} was also reported by Kroto and, as expected, it contained five peaks (see Fig. 3 - 8). The proposed football-shaped C_{70} fullerene possesses five sets of inequivalent carbon atoms in a ratio of 10 : 10 : 20 : 20 : 10. This is precisely the ratio of the line intensities observed in the ^{13}C NMR spectrum.

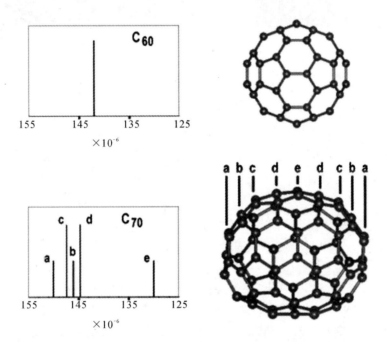

Fig. 3 - 8 Idealized ^{13}C NMR spectra and structural drawings of C_{60} (top) and C_{70} (bottom).
In C_{60}, all carbon atoms are identical and a single ^{13}C NMR peak is observed. In C_{70}, there are five sets of inequivalent carbon atoms (labelled a-e), giving rise to five ^{13}C NMR signals

3.5.2 Single crystal X-ray structure determinations

Definitive proof for the structure of C_{60} came in 1991 when Joel Hawkins (University of California-Berkeley) synthesized and crystallized an osmium derivative of C_{60} (see Fig. 3 - 9), $(C_{60})O_sO_4$ (4-tert-butylpyridine)$_2$ (Hawkins, 1991). Single crystal X-ray analysis of this compound yielded the first atomic-resolution picture of the carbon framework of C_{60} (see Fig. 3 - 10). (Note: Because of its high symmetry, C_{60} itself is orientationally disordered in the solid state. In fact, at ambient temperature, the C_{60} molecules rotate rapidly in the solid state. Derivatization with O_sO_4 breaks the nearly spherical symmetry of C_{60}, allowing it to crystallize with orientational order.)

Fig. 3-9 C_{60} fullerene in crystalline form

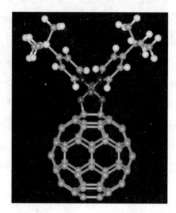

Fig. 3-10 X-ray crystal structure of $(C_{60})O_sO_4$ (4-tert-butylpyridine) (see inside back cover)

Note: The carbon atoms are green, hydrogen atoms are white, nitrogen atoms are dark blue, osmium atom is magenta, and oxygen atoms are red in the ball-and-stick representation

Osmylation occurs selectively across a 6-6 ring junction rather than a 6-5 ring junction. Not surprisingly, bond distances in the osmylated portion of C_{60} are dramatically affected (C_1-C_2, C_1-C_3, C_1-C_4, C_2-C_5, and C_2-C_6 bond lengths average 1.55 Å, comparable to normal C − C single bonds). However, the remainder of the C_{60} structure is not significantly perturbed by the O_sO_4 unit. Interestingly, there are statistically-different average bond lengths for the 6-6 and 6-5 ring fusions. Excluding bonds to C_1 and C_2 (the osmylated carbons), the average C − C bond lengths are 1.386(9) Å for 6-6 ring fusions and 1.434 Å for 6-5 ring fusions. (Note: these average bond lengths have subsequently been confirmed in low-temperature neutron powder diffraction and gas-phase electron diffraction studies of C_{60} itself.) Hence, the bonding in C_{60} is not completely delocalized as it is in graphite. Rather, the dominant resonance structure (see Fig. 3-11) is one in which the double bonds are located "exocyclic" to the five-membered rings and between the six-membered rings. This explains the regiochemistry of the osmylation reaction; it occurs preferentially on the more electron-rich 6 − 6 ring fusion.

Structural proof for C_{70} also came in 1991 when Balch and coworkers succeeded in crystallizing and obtaining the single crystal X-ray structure of $(C_{70})Ir(CO)(Cl)(PPh_3)_2$ (see Fig. 3-12). (Balch, 1991, JACS) This adduct is particularly interesting from the point of view of regiochemistry because there are four types of 6-6 ring fusions in C_{70} as well as four 6-5 fusions. The metal selectively positions itself across one of the 6-6 fusions near the elongated end of the football-shaped molecule.

Fig. 3-11 Dominant resonance structure for C_{60}, in which formal double bonds are located at 6-6 ring fusions

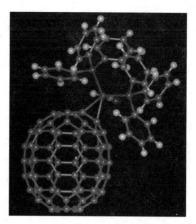

Fig. 3-12 X-ray crystal structure of $(C_{70})Ir(CO)(Cl)(PPh_3)_2$
(see inside back cover)

Note: The carbon atoms are green, chlorine atom is yellow, hydrogen atoms are white, iridium atom is light blue, oxygen atom is red, and phosphorus atoms are magneta in the ball-and-stick representation

In general, metal binding is accompanied by local distortion so that the two carbon atoms involved in the coordination are pulled out from the fullerene surface. Simple geometric considerations show that the 6-6 ring fusion at the elongated end is the most accessible. The other 6-6 ring fusions all have a more flattened local structure and would require much larger distortions to accommodate metal coordination.

3.6 Chemistry of fullerenes

3.6.1 Typical reactions of fullerenes

For hollow objects like the fullerenes, a general distinction has to be made between outside and inside reactivity. Modifications to the outside are termedexohedral functionalization and those to the inside are endohedral. Both variants are observed for the fullerenes.

Classical fullerene chemistry deals with exohedral functionalization by one or more groups attached to the carbon atoms. Endohedral chemistry, on the other hand, studies compounds consisting of atoms or small molecules included in the cavity within the fullerene cage. The exohedral processes may further be divided into covalent and noncovalent interactions with the reaction partner. (see Fig. 3-13 and Fig. 3-14)

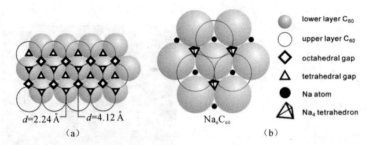

Fig. 3-13 The structure of C_{60}

(a) The structure of the cubic face-centered lattice of C_{60} with intercalation sites; (b) example of the intercalation compounds Na_6C_{60} with Na_4-tetrahedra occupying the octahedral gaps of the fullerene lattice

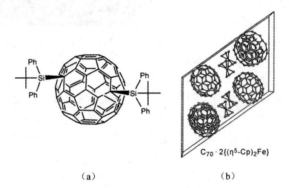

Fig. 3-14 C_{60} and C_{70} compounds

(a) Product of the 1,6-addition of a silicon compound to C_{60}; (b) co-crystallizate of C_{70} and ferrocene. An attack by the latter does not occur as its reduction power is too low

3.6.2 Metallofullerenes

As seen before, the inside of the fullerene cages is empty, so it is self-suggesting to try to fill this cavity with atoms or molecules. Indeed the isolation of so-called endohedral fullerenes succeeded quite soon after the discovery of fullerenes themselves. In endohedral species, single or even several atoms are situated inside the carbon cage, so it is purely topological circumstances that prevent them from breaking out of these compounds.

The preparation of metallofullerenes can be achieved in different ways. The methods most commonly used today are the arc-evaporation of impregnated graphitic rods and the laser-evaporation process (also called the laser-oven method). For the latter a rotating target of graphite and metal oxide (cemented with a pitch binder) is placed in an oven heated to about 1,200 ℃ and irradiated with a doubled-frequency Nd:YAG laser under a stream of argon. The resulting fullerenes and metallofullerenes are swept along with the gas current and precipitated in the cooler zones at the end of the quartz tube. A temperature of not less than 800 ℃ is required for this process as neither empty nor filled fullerenes are formed below that value.

The arc-method uses the same procedure as successfully employed to produce empty fullerenes, with the sole difference that the anodes of graphite evaporated in a classical arc-apparatus are impregnated with metal oxides or carbides. (In the graphitic rods treated with metal oxides, the respective carbides will be generated too when heated to $>1,600\,℃$.) Besides empty species, the soot deposited on the cooler reactor walls contains various metallofullerenes. Endofullerenes like lithium end of ullerene ($Li@C_{60}$) can further be obtained by ion implantation techniques. High-energy ions of the desired element are targeted at a thin film of fullerenes. However, the produced amounts are small and accordingly the analysis of products is complicated. Apart from the expected $M@C_{60}$, the examination of soot containing metallofullerenes also reveals carbon cages of unusual sizes. $La@C_{74}$ and $La@C_{82}$ are observed upon evaporation of a graphitic sample treated with La_2O_3. The C_{82}-metallofullerene exhibits the highest stability of all. It may, as an extract in toluene, even be stored under air without decomposing. $La@C_{60}$, on the other hand, cannot be isolated due to its instability. (see Fig.3-15 to Fig.3-18)

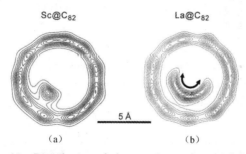

Fig. 3-15 Distribution of electron density in endofullerenes.
(a) Scandium remains fixed at its position out of the cage's center; (b) Lanthanum exerts a motion within the cage

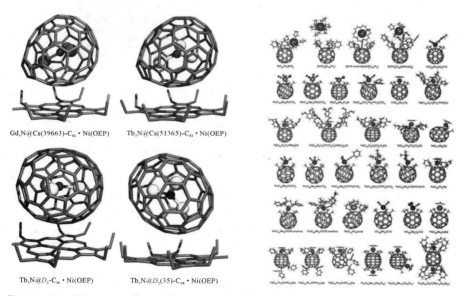

Fig. 3-16 X-ray crystallographic structures of metal-penta(organo) fullerene complexes

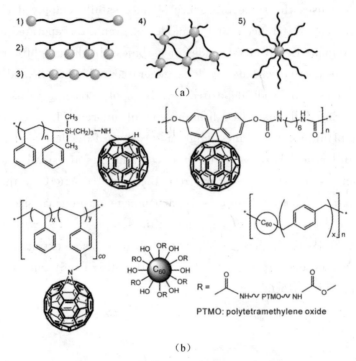

Fig. 3-17　Fullerene-polymer composites
(a) Different kinds of polymer materials that may be obtained from different modes of fullerene incorporation;　(b) Examples of fullerene-polymer composites

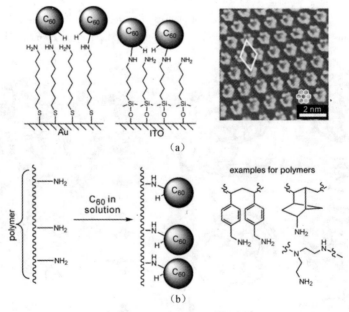

Fig. 3-18　Immobilization of C_{60} on surfaces
(a) Covalent linking to gold or to ITO and noncovalent attachment to gold modified with coronene;
(b) Attachment of C_{60} to polymers by nucleophilic addition of terminal amino groups

3.7 The doped treatment and application of fullerenes

Fullerenes are stable, but not totally unreactive. The sp^2-hybridized carbon atoms, which are at their energy minimum in planar graphite, must be bent to form the closed sphere or tube, which produces angle strain. The characteristic reaction of fullerenes is electrophilic addition at 6,6-double bonds, which reduces angle strain by changing sp^2-hybridized carbons into sp^3-hybridized ones. The change in hybridized orbitals causes the bond angles to decrease from about 120 degrees in the sp^2 orbitals to about 109.5 degrees in the sp^3 orbitals. This decrease in bond angles allows for the bonds to bend less when closing the sphere or tube, and thus, the molecule becomes more stable.

Other atoms can be trapped inside fullerenes to form inclusion compounds known as endohedral fullerenes. An unusual example is the egg shaped fullerene $Tb_3N@C_{84}$, which violates the isolated pentagon rule. Recent evidence for a meteor impact at the end of the Permian period was found by analyzing noble gases so preserved. Metallofullerene-based inoculates using the rhonditic steel processes are beginning production as one of the first commercially-viable uses of buckyballs.

3.7.1 Solubility

Fullerenes are sparingly soluble in many solvents. Common solvents for the fullerenes include aromatics, such as toluene, and others like carbon disulfide. Solutions of pure buckminsterfullerene have a deep purple color. Solutions of C_{70} are a reddish brown. The higher fullerenes C_{76} to C_{84} have a variety of colors. C_{76} has two optical forms, while other higher fullerenes have several structural isomers. Fullerenes are the only known allotrope of carbon that can be dissolved in common solvents at room temperature.

Some fullerene structures are not soluble because they have a small band gap between the ground and excited states. These include the small fullerenes C_{28}, C_{36} and C_{50}. The C_{72} structure is also in this class, but the endohedral version with a trapped lanthanide-group atom is soluble due to the interaction of the metal atom and the electronic states of the fullerene. Researchers had originally been puzzled by C_{72} being absent in fullerene plasma-generated soot extract, but found in endohedral samples. Small band gap fullerenes are highly reactive and bind to other fullerenes or to soot particles.

Solvents that are able to dissolve buckminsterfullerene (C_{60}) are listed below in order from highest solubility. The value in parentheses is the approximate saturated concentration.

(1) 1-chloronaphthalene (51 mg/mL).
(2) 1-methylnaphthalene (33 mg/mL).
(3) 1,2-dichlorobenzene (24 mg/mL).
(4) 1,2,4-trimethylbenzene (18 mg/mL).
(5) tetrahydronaphthalene (16 mg/mL).
(6) carbon disulfide (8 mg/mL).

(7) 1,2,3-tribromopropane (8 mg/mL).
(8) bromoform (5 mg/mL).
(9) cumene (4 mg/mL).
(10) toluene (3 mg/mL).
(11) benzene (1.5 mg/mL).
(12) cyclohexane (1.2 mg/mL).
(13) carbon tetrachloride (0.4 mg/mL).
(14) chloroform (0.25 mg/mL).
(15) n-hexane (0.046 mg/mL).
(16) tetrahydrofuran (0.006 mg/mL).
(17) acetonitrile (0.004 mg/mL).
(18) methanol (0.000,04 mg/mL).
(19) water (1.3×10^{-11} mg/mL).

Solubility of C_{60} in some solvents shows unusual behaviour due to existence of solvate phases (analogues of crystallohydrates). For example, solubility of C_{60} in benzene solution shows maximum at about 313 K. Crystallization from benzene solution at temperatures below maximum results in formation of triclinic solid solvate with four benzene molecules $C_{60} \cdot 4C_6H_6$ which is rather unstable in air. Out of solution, this structure decomposes into usual fcc C_{60} in few minutes' time. At temperatures above solubility maximum the solvate is not stable even when immersed in saturated solution and melts with formation of fcc C_{60}. Crystallization at temperatures above the solubility maximum results in formation of pure fcc C_{60}. Large millimetre size crystals of C_{60} and C_{70} can be grown from solution both for solvates and for pure fullerenes.

3.7.2 Superconductivity

After the synthesis of macroscopic amounts of fullerenes, their physical properties could be investigated. Very soon A. F. Haddon et al. found that intercalation of alkali-metal atoms in solid C_{60} leads to metallic behavior. In 1991, it was revealed that potassium-doped C_{60} becomes superconducting at 18K. This was the highest transition temperature for a molecular superconductor. Since then, superconductivity has been reported in fullerene doped with various other alkali metals. It has been shown that the superconducting transition temperature in alkaline-metal-doped fullerene increases with the unit-cell volume V. As cesium forms the largest alkali ion, cesium-doped fullerene is an important material in this family. Recently, superconductivity at 38 K has been reported in bulk Cs_3C_{60}, but only under applied pressure. The highest superconducting transition temperature of 33 K at ambient pressure is reported for $Cs_2R_bC_{60}$.

The increase of transition temperature with the unit-cell volume had been believed to be evidence for the BCS mechanism of C_{60} solid superconductivity, because inter C_{60} separation can be related to an increase in the density of states on the Fermi level, $N(\varepsilon_F)$. Therefore, there have been many efforts to increase the interfullerene separation, in particular, intercalating neutral molecules into the A_3C_{60} lattice to increase the interfullerene spacing while the valence of C_{60} is kept unchanged. However, this ammoniation technique has revealed a new aspect of fullerene intercalation compounds: the Mott-Hubbard transition and the correlation between the orientation/orbital order of C_{60} molecules and the magnetic structure.

The C_{60} molecules compose a solid of weakly bound molecules. The fullerites are therefore molecular solids, in which the molecular properties still survive. The discrete levels of a free C_{60} molecule are only weakly broadened in the solid, which leads to a set of essentially nonoverlapping bands with a narrow width of about 0.5 eV. For an undoped C_{60} solid, the 5-fold h_u band is the HOMO level, and the 3-fold t_{1u} band is the empty LUMO level, and this system is a band insulator. But when the C_{60} solid is doped with metal atoms, the metal atoms give electrons to the t_{1u} band or the upper 3-fold t_{1g} band. This partial electron occupation of the band leads to sometimes metallic behavior. However, A_4C_{60} is an insulator, although the t_{1u} band is only partially filled and it should be a metal according to band theory. This unpredicted behavior may be explained by the Jahn-Teller effect, where spontaneous deformations of high-symmetry molecules induce the splitting of degenerate levels to gain the electronic energy. The Jahn-Teller type electron-phonon interaction is strong enough in C_{60} solids to destroy the band picture for particular valence states.

A narrow band or strongly correlated electronic system and degenerated ground states are important points to understand in explaining superconductivity in fullerene solids. When the inter-electron repulsion U is greater than the bandwidth, an insulating localized electron ground state is produced in the simple Mott-Hubbard model. This explains the absence of superconductivity at ambient pressure in cesium-doped C_{60} solids. Electron-correlation-driven localization of the t1u electrons exceeds the critical value, leading to the Mott insulator. The application of high pressure decreases the interfullerene spacing, therefore cesium-doped C_{60} solids turn to metallic and superconducting.

A fully developed theory of C_{60} solids superconductivity is still lacking, but it has been widely accepted that strong electronic correlations and the Jahn-Teller electron-phonon coupling produce local electron-pairings that show a high transition temperature close to the insulator-metal transition.

3.8 Possible applications of fullerenes and their derivatives

In the last few years, scientists have employed the different methodologies for the functionalization of fullerenes, as well as the different types of supermolecular interactions to build a variety of very interesting molecular architectures with efficient applications.

In the next section, some of the most promising fields of application of fullerene-containing systems will be summarized. The inherent photo- and electrochemical properties of fullerenes have played an active role in the performance of the system.

Possible applications of fullerenes and their derivatives (see Fig. 3-19 and Fig. 3-20):

(1) Solar cells.
(2) Composite materials with interesting electronic properties.
(3) Derivatives of fullerenes for photodynamic tumor therapy.
(4) Chemical sensors.
(5) Endofullerenes as contrasting agent in MRI with reduced side effects.
(6) Superconductors.
(7) Donor-acceptor systems.
1) Polyads.
2) Dyadscontaining nonphotoactive electron donors.
3) Dyadscontaining photoactive electron donors.

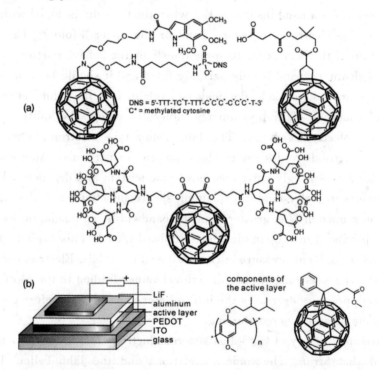

Fig. 3-19 Functionalized fullerene materials may be employed in biological or electronic applications
(a) Examples of water-soluble derivatives of fullerene that might be suitable for medical uses;
(b) Solar cells with their active layer consisting of a C_{60}-composite

The organic functionalization of C_{60} has produced a wide range of derivatives, which retain the basic properties of pristine fullerene. Among the many possible reactions available, cycloadditions have been most widely used, along with cyclopropanation (nucleophilic) reactions.

The products have now improved the solubility and processability and can be used in several applications including electron-transfer reactions, liquid crystals, polymers, dendrimers, and solar cells. The continuous evolution of fullerene science and technology, accompanying the progress obtained in the functionalization chemistry, has led to the production of more and more compounds that open new horizons in the potential applications of these fascinating molecules.

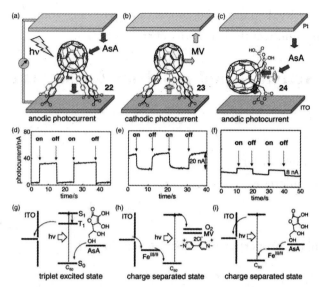

Fig. 3-20 Photocurrent generation properties of fullerene pentacarboxylic acids 22-24
(a)-(c) Molecular structures and orientation with direction of photocurrent;
(d)-(f) On-off profiles of photocurrent generation (positive current: anodic; negative current: cathodic);
(g)-(i) Difference in excited state and mechanism for photocurrent generation

When fullerenes were first discovered, nobody thought of real commercial applications of the new material — the structure simply seemed too exotic to become available in macroscopic amounts.

However, as the methods of producing C_{60} have continually been improved to the present day, the price per gram has been reduced to about 100 (highly purified lab quality), so in fact it is obtainable for various applications. Some of these shall be explained in short.

On one hand, C_{60} and its derivatives are considered promising materials for biological uses. The farthest progress has been made in experiments on its employment as sensitizer for the photochemical generation of singlet oxygen. The latter shall serve to the directed degradation of DNA or to the destruction of tumor tissue (photodynamic therapy). On the other hand, it is the electronic properties of the cage-like molecule that are of interest for applications as redox-active substance or as surface film.

Hydrofullerenes with various degrees of hydrogenation are candidate materials for hydrogen storage.

$C_{60}H_{36}$, for instance, has a hydrogen storing capacity of 4.8% of its own mass. This value may be below the 6.5% required for an economical application, but still it is a considerable improvement compared to other materials.

Hydrogenated fullerenes may also be useful in lithium ion accumulators because they significantly prolong the batteries lifetime.

Summing up, a wide repertoire of possible applications has been developed for the fullerenes, and most of all for C_{60}. But despite a massive decrease of prices for fullerene soot, the cost for the starting material still is the biggest impediment to a profitable use of the outstanding properties. Real commercial applications will only come into reach when C_{60} can be produced at prices significantly below the current level.

Activities and problems for students

Activities.

(1) Is C_{60} the smallest fullerene?
(2) How to dope fullerenes?

Problems.

What are the unique properties of carbon nanopeapods?

Chapter 4   Carbon Nanotube

4.1   History of carbon nanotube

In 1980 we knew of only three forms of carbon, namely diamond, graphite, and amorphous carbon. Today we know there is a whole family of other forms of carbon. The first to be discovered was the hollow, cage-like Buckminsterfullerene molecule — also known as the buckyball, or the C_{60} fullerene. There are now thirty or more forms of fullerenes, and also an extended family of linear molecules, carbon nanotubes. C_{60} is the first spherical carbon molecule, with carbon atoms arranged in a soccer ball shape. In the structure there are 60 carbon atoms and a number of five-membered rings isolated by six-membered rings. The second, slightly elongated, spherical carbon molecule in the same group resembles a rugby ball, has seventy carbon atoms and is known as C_{70}. C_{70}'s structure has extra six-membered carbon rings, but there are also a large number of other potential structures containing the same number of carbon atoms. Their particular shapes depend on whether five-membered rings are isolated or not, or whether seven-membered rings are present. Many other forms of fullerenes up to and beyond C_{120} have been characterized, and it is possible to make other fullerene structures with five-membered rings in different positions and sometimes adjoining one another.

The important fact for nanotechnology is that useful dopant atoms can be placed inside the hollow fullerene ball. Atoms contained within the fullerene are said to be endohedral. Of course they can also be bonded to fullerenes outside the ball as salts, if the fullerene can gain electrons.

Endohedral fullerenes can be produced in which metal atoms are captured within the fullerene cages. Theory shows that the maximum electrical conductivity is to be expected for endohedral metal atoms, which will transfer three electrons to the fullerene. Fullerenes can be dispersed on the surface as a monolayer. That is, there is only one layer of molecules, and they are said to be mono dispersed. Provided fullerenes can be placed in very specific locations, they may be aligned to form a fullerene wire. Systems with appropriate material inside the fullerene ball are conducting and are of particular interest because they can be deposited to produce bead-like conducting circuits. Combining endohedrally doped structures with non-doped structures changes the actual composition of a fullerene wire, so that it may be tailored in-situ during patterning. Hence within a single wire, insulating and conducting regions may be precisely defined. One-dimensional junction engineering becomes realistic with fullerenes.

Possibly more important than fullerenes are carbon nanotubes which are related to graphite. The molecular structure of graphite resembles stacked, one-atom-thick sheets of chicken wire — a planar network of interconnected hexagonal rings of carbon atoms. In conventional graphite, the sheets of carbon are stacked on top of one another, allowing them to easily slide over each other. That is why graphite is not hard, but it feels greasy, and can be used as a lubricant. When graphene sheets are rolled into a cylinder and their edges joined, they form carbon nanotubes (CNTs). Only the tangents of the graphitic planes come into contact with each other, and hence their properties are more like those of a molecule.

CNTs come in a variety of diameters, lengths and functional group content. CNTs today are available for industrial applications in bulk quantities up metric ton quantities from cheap tubes. Several CNT manufacturers have >100 ton per year production capacity for multi walled nanotubes.

A nanotube may consist of one tube of graphite, a one-atom thick single-wall nanotube or a number of concentric tubes called multiwalled nanotubes. When viewed with a transmission electron microscope these tubes appear as planes. Whereas single-walled nanotubes appear as two planes, in multi-walled nanotubes more than two planes are observed, and can be seen as a series of parallel lines. There are different types of CNTs, because the graphitic sheets can be rolled in different ways. The three types of CNTs are zigzag, armchair, and chiral. It is possible to recognize zigzag, armchair, and chiral CNTs just by following the pattern across the diameter of the tubes, and analyzing their cross-sectional structure.

Multi-walled nanotubes can come in an even more complex array of forms, because each concentric single-walled nanotube can have different structures, and hence there are a variety of sequential arrangements. The simplest sequence is when concentric layers are identical but different in diameter. However, mixed variants are possible, consisting of two or more types of concentric CNTs arranged in different orders. These can have either regular layering or random layering. The structure of the nanotube influences its properties — including electrical and thermal conductivity, density, and lattice structure. Both type and diameter are important. The wider the diameter of the nanotube, the more it behaves like graphite. The narrower the diameter of the nanotube, the more its intrinsic properties depends upon its specific type.

4.2　The discovery of carbon nanotube

A 2006 editorial written by Marc Monthioux and Vladimir Kuznetsov in the journal Carbon described the interesting and often misstated origin of the carbon nanotube. A large percentage of academic and popular literature attributes the discovery of hollow, nanometer-size tubes composed of graphitic carbon to Sumio Iijima of NEC in 1991.

In 1952, L. V. Radushkevich and V. M. Lukyanovich published clear images of 50 nanometer diameter tubes made of carbon in the Soviet Journal of Physical Chemistry. This discovery was largely unnoticed, as the article was published in the Russian language, and Western scientists' access to Soviet press was limited during the Cold War. It is likely that carbon nanotubes were produced before this date, but the invention of the transmission electron microscope (TEM) allowed direct visualization of these structures.

Carbon nanotubes have been produced and observed under a variety of conditions prior to 1991. A paper by Oberlin, Endo and Koyama published in 1976 clearly showed hollow carbon fibers with nanometer-scale diameters using a vapor-growth technique. Additionally, the authors show a TEM image of a nanotube consisting of a single wall of graphene. Later, Endo has referred to this image as a single-walled nanotube.

In 1979 John Abrahamson presented evidence of carbon nanotubes at the 14th Biennial Conference of Carbon at Pennsylvania State University. The conference paper described carbon nanotubes as carbon fibers which were produced on carbon anodes during arc discharge. A characterization of these fibers was given as well as hypotheses for their growth in a nitrogen atmosphere at low pressures.

In 1981 a group of Soviet scientists published the results of chemical and structural characterization of carbon nanoparticles produced by a thermo catalytical disproportionation of carbon monoxide. Using TEM images and XRD patterns, the authors suggested that their "carbon multi-layer tubular crystals" were formed by rolling graphene layers into cylinders. They speculated that by rolling graphene layers into a cylinder, many different arrangements of graphene hexagonal nets are possible. They suggested two possibilities of such arrangements: circular arrangement (armchair nanotube) and a spiral, helical arrangement (chiral tube).

In 1987, Howard G. Tennent of Hyperion Catalysis was issued a U.S. patent for the production of "cylindrical discrete carbon fibrils" with a "constant diameter between about 3.5 nm and about 70 nm..., length 102 times the diameter, and an outer region of multiple essentially continuous layers of ordered carbon atoms and a distinct inner core..."

Iijima's discovery of multi-walled carbon nanotubes in the insoluble material of arc-burned graphite rods in 1991 and Mintmire, Dunlap, and White's independent prediction that if single-walled carbon nanotubes could be made, then they would exhibit remarkable conducting properties helped create the initial buzz that is now associated with carbon nanotubes. Nanotube research accelerated greatly following the independent discoveries by Bethune at IBM and Iijima at NEC of single-walled carbon nanotubes and methods to specifically produce them by adding transition-metal catalysts to the carbon in an arc discharge. The arc discharge technique was well-known to produce the famed Buckminster-fullerene on a preparative scale, and these results appeared to extend the run of accidental discoveries relating to fullerenes. The original observation of fullerenes in mass spectrometry was not anticipated, and the first mass-production technique by Krätschmer and Huffman was used for several years before realizing that it produced fullerenes.

The discovery of nanotubes remains a contentious issue, especially because several scientists involved in the research could be likely candidates for the Nobel Prize. Many believe that Iijima's report in 1991 is of particular importance because it brought carbon nanotubes into the awareness of the scientific community as a whole. See the reference for a review of the history of the discovery of carbon nanotubes.

Similar to the matter of nanotube discovery is the question of what is the thinnest possible carbon nanotube. Possible candidates are: nanotubes of about 0.40 nm diameter have been reported in 2000; however, they are not free standing, but enclosed in zeolite crystals or are innermost shells of the multi-wall nanotubes. Later, inner shells of MWCNTs of only 0.3 nm in diameter have been reported. The thinnest free-standing nanotube, by September 2003, has diameter of 0.43 nm.

Carbon nanotubes (CNTs) were observed by Sumio Iijima in 1991 (see Fig. 4 - 1). CNTs are extended tubes of rolled graphene sheets. There are two types of CNT: single-walled (one tube) or multi-walled (several concentric tubes). Both of these are typically a few nanometres in diameter and several micrometres to centimetres long. CNTs have assumed an important role in the context of nanomaterials, because of their novel chemical and physical properties. They are mechanically very strong (their Young's modulus is over 1 TPa, making CNTs as stiff as diamond), flexible (about their axis), and can conduct electricity extremely well (the helicity of the graphene sheet determines whether the CNT is a semiconductor or metallic). All of these remarkable properties give CNTs a range of potential applications: for example, in reinforced composites, sensors, nanoelectronics and display devices.

CNTs are now available commercially in limited quantities. They can be grown by several techniques. However, the selective and uniform production of CNTs with specific dimensions and physical properties is yet to be achieved. The potential similarity in size and shape between CNTs and asbestos fibres has led to concerns about their safety.

Carbon nanotubes (CNTs) are allotropes of carbon with a cylindrical nanostructure. Nanotubes have been constructed with length-to-diameter ratio of up to 28,000,000 : 1, which is significantly larger than any other material. These cylindrical carbon molecules have novel properties that make them potentially useful in many applications in nanotechnology, electronics, optics and other fields of materials science, as well as potential uses in architectural fields. They exhibit extraordinary strength and unique electrical properties, and are efficient conductors of heat. Their final usage, however, may be limited by their potential toxicity.

Nanotubes are members of the fullerene structural family, which also includes the spherical buckyballs. The ends of a nanotube might be capped with a hemisphere of the buckyball structure. Their name is derived from their size, since the diameter of a nanotube is on the order of a few nanometers (approximately 1/50,000th of the width of a human hair), while they can be up to several millimeters in length (as of 2,008). Nanotubes are categorized as single-walled carbon nanotubes (SWCNTs) and multi-walled carbon nanotubes (MWCNTs).

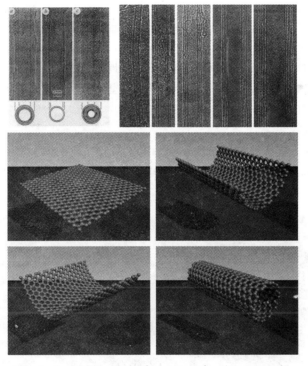

Fig. 4 - 1   TEM and the formation of carbon nanotube

The nature of the bonding of a nanotube is described by applied quantum chemistry, specifically, orbital hybridization. The chemical bonding of nanotubes is composed entirely of sp^2 bonds, similar to those of graphite. This bonding structure, which is stronger than the sp^3 bonds found in diamonds, provides the molecules with their unique strength. Nanotubes naturally align themselves into "ropes" held together by van der Waals forces (see Fig. 4 - 2). Under high pressure, nanotubes can merge together, trading some sp^2 bonds for sp^3 bonds, giving the possibility of producing strong, unlimited-length wires through high-pressure nanotube linking.

 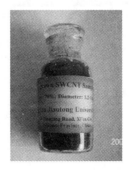

Fig. 4 - 2   The samples of carbon nanotubes (right: produced by us)

4.3 The type of carbon nanotube

Since the discovery of fullerene in 1985 and carbon nanotubes (CNTs) in 1991, there has been a worldwide research on novel carbon nano-materials because of their unique physical properties. Carbon nanotubes have attracted great attention in the past decade due to their interesting properties and potential applications.

At present, CNTs have two dominant types:
(1) crystallized CNTs;
(2) amorphous CNTs (ACNTs) (see Fig.4 - 3 and Fig.4 - 4).

Generally, crystallized CNTs include single-walled CNTs (SWCNTs) and multi-walled CNTs (MWCNTs).

Fig. 4 - 3 Optical photograph of cloth-like ACNTs

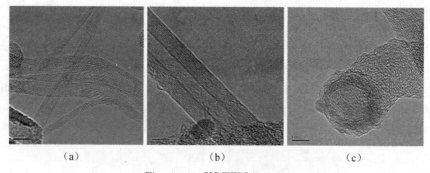

(a)　　　　　　　　(b)　　　　　　　　(c)

Fig. 4 - 4 HRTEM images
(a) SWCNTs;　(b) MWCNTs;　(c) ACNTs

4.4 The synthesis of carbon nanotube

There are a number of methods of making CNTs and fullerenes. Fullerenes were first observed after vaporizing graphite with a short-pulse, high-power laser, however this was not a practical method for making large quantities. CNTs have probably been around for a lot longer than was first realized, and may have been made during various carbon combustion and vapor deposition processes, but electron microscopy at that time was not advanced enough to distinguish them from other types of tubes. The first method for producing CNTs and fullerenes in reasonable quantities was by applying an electric current across two carbonaceous electrodes in an inert gas atmosphere. This method is called plasma arcing. It involves the evaporation of one electrode as cations followed by deposition at the other electrode. This plasma-based process is analogous to the more familiar electroplating process in a liquid medium. Fullerenes and CNTs are formed by plasma arcing of carbonaceous materials, particularly graphite. The fullerenes appear in the soot that is formed, while the CNTs are deposited on the opposing electrode. Another method of nanotube synthesis involves plasma arcing in the presence of cobalt with a 3% or greater concentration. As noted above, the nanotube product is a compact cathode deposit of rod like morphology. However when cobalt is added as a catalyst, the nature of the product changes to a web, with strands of 1 mm or so thickness that stretch from the cathode to the walls of the reaction vessel. The mechanism by which cobalt changes this process is unclear, however one possibility is that such metals affect the local electric fields and hence the formation of the five-membered rings.

4.4.1 Arc method

The carbon arc discharge method, initially used for producing C_{60} fullerenes, is the most common and perhaps easiest way to produce CNTs, as it is rather simple (see Fig. 4-5). However, it is a technique that produces a complex mixture of components, and requires further purification to separate the CNTs from the soot and the residual catalytic metals present in the crude product. This method creates CNTs through arc-vaporization of two carbon rods placed end to end, separated by approximately 1 mm, in an enclosure that is usually filled with inert gas at low pressure. Recent investigations have shown that it is also possible to create CNTs with the arc method in liquid nitrogen. A direct current of 50 A to 100 A, driven by a potential difference of approximately 20 V, creates a high temperature discharge between the two electrodes. The discharge vaporizes the surface of one of the carbon electrodes, and forms a small rod-shaped deposit on the other electrode. Producing CNTs in high yield depends on the uniformity of the plasma arc, and the temperature of the deposit forming on the carbon electrode.

However the first macroscopic production of carbon nanotubes was made in 1992 by two researchers at NEC's Fundamental Research Laboratory. The method used was the same as in 1991. During this process, the carbon contained in the negative electrode sublimates because of the high discharge temperatures.

Because nanotubes were initially discovered using this technique, it has been the most widely-used method of nanotube synthesis.

The yield for this method is up to 30 percent by weight and it produces both single- and multi-walled nanotubes with lengths of up to 50 micrometers.

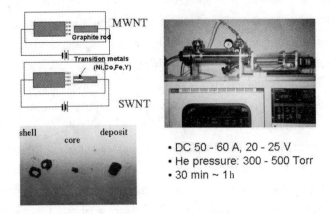

Fig. 4-5　Arc-discharge system for the synthesis of carbon nanotube

> **Temperature-controlled arc discharge.**

Since the discovery of single-walled carbon nanotubes in 1993, the large scale and high purity synthesis of SWCNTs has become an important research task. SWCNTs have many potential applications such as in nanodevices, field emission display, molecular computers, and energy storage media. Significant progress has recently been achieved in all three of the main synthesis methods: arc discharge, laser ablation and chemical vapor deposition (CVD). Though arc discharge is a suitable method for production of SWCNTs, high quality SWCNTs can only be found in the collaret around the cathode, resulting in very low productivity. In order to raise the production, Liu et al. used a semi-continuous hydrogen arc discharge method to prepare SWCNTs, and obtained 2 g/h SWCNT ropes, the proportion of SWCNTs in the ropes is about 30%, the diameter of SWCNTs is about 1.72 nm. Ando et al. developed a DC arc plasma jet method and the highest yield was 1.24 g/min and the purity was about 50%. Morio Takizawa et al. studied the effect of environmental temperature for synthesizing SWCNTs by arc vaporization. They conducted the arc discharge in a quartz tube 50 mm in diameter and 300 mm in length, and controlled temperature outside the tube. Because the temperature in the arc position is very high and the arc space in the tube is too small, the real temperature in the quartz tube cannot be known. The content of SWCNTs produced by conventional arc discharge is about 30 wt.% (wt. is short for weight) in "collar" soot and ~10 wt.% in "wall" soot, which is not evenly distributed over the entire wall. At present, the main disadvantages with conventional arc discharging are low yield and purity.

In order to improve this, we have modified the conventional arc discharge apparatus. Our apparatus can freely control the temperature of the chamber in the range from room temperature to 900 ℃. Furthermore, the modified arc discharge furnace can prepare SWCNTs in large scale (>45 g/h) and high purity (95% after purification) on the wall of the chamber, comparable to the HiPco laser ablation method.

Chapter 4 Carbon Nanotube

Recently, amorphous carbon nanotubes (ACNTs) have attracted great attention to many researchers. Amorphous carbon nanotube with diameter in the range of 3 - 60 nm is different from crystalline carbon nanotubes. The walls of the ACNTs are composed of many carbon clusters whose characteristic is of short-distance order and long-distance disorder. Therefore, properties of the ACNTs are different from single-walled and multi-walled carbon nanotubes. It has been reported that ACNTs can be synthesized by CVD recently. We have modified the traditional arc discharge apparatus with a temperature-controlled system. The temperature can be maintained at designed values during the arcing process. Both crystalline carbon nanotubes and amorphous carbon nanotoubes can be formed not only on the cathode but also on all the walls of the vacuum chamber in a large scale.

Our experimental apparatus was designed by Professor Liu Yongning et al. (see Fig. 4 - 6). The vacuum chamber is $\phi 300 \times 400$ mm^2. And there is a heating apparatus mounted outside the chamber to control the temperature inside the chamber. There are six $\phi 6 \times 100$ mm^2 graphite rode anode with a $\phi 4 \times 60$ mm^2 hole drilled in them which is filled with a 19 : 1 (weight ratio) mixture of Fe-Ni-Mg (2 : 1 : 2 wt.%) powders. The six anodes are mounted at an equal distance from each other on a wheel which can be rotated in order to change the active anode. In our experiments: an arc was generated between the anode and the pure graphite cathode at a current of 60 A in a 400 Torr of static helium atmosphere. The distance between the electrodes was maintained at 2 mm by continuously feeding the cathode throughout the arc process. We observed a cloth-like soot had formed on the entire inner wall of the chamber and, in general, an 80 mm anode rod is used up in 5 min and 5.3 g of soot can be collected(see Fig.4 - 7 to Fig.4 - 13).

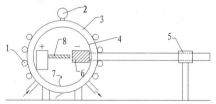

Fig. 4 - 6 Above are optical and schematic drawing of arc discharge apparatus which is capable of carrying out experiments at different temperatures

1 - Water-cooling system; 2 - Vacuum pressure-meter; 3 - Vacuum chamber; 4 - Temperature controlled apparatus;
5 - Electrode feeding system; 6 - Moving cathode; 7 - Thermal couple; 8 - Fixed and rotated anode

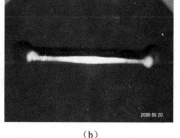

(a)   (b)

Fig. 4 - 7 Optical photos
(a) SWCNTs; (b) ACNTs

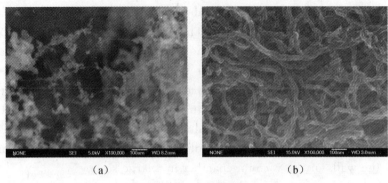

Fig. 4-8 SEM images
(a) SWCNTs; (b) ACNTs

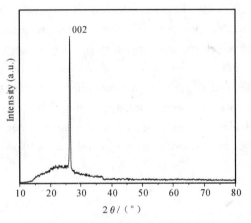

Fig. 4-9 XRD pattern of SWCNTs

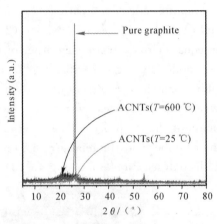

Fig. 4-10 XRD patterns of pure graphite and ACNTs

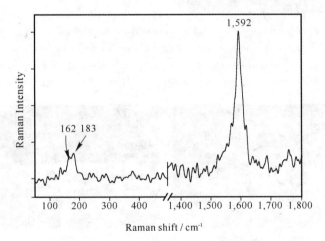

Fig. 4-11 Raman pattern of SWCNTs

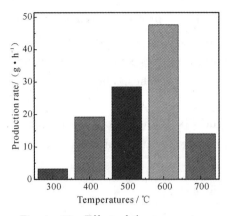
Fig. 4-12 Effect of the temperature on production rate of the SWCNTs

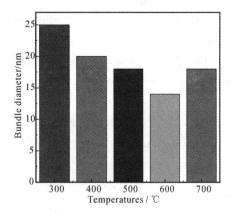
Fig. 4-13 Effect of the temperature on bundle average diameter of the SWCNTs

4.4.2 Laser method

In the laser ablation process, apulsed laser vaporizes a graphite target in a high-temperature reactor while an inert gas is bled into the chamber. Nanotubes develop on the cooler surfaces of the reactor as the vaporized carbon condenses. A water-cooled surface may be included in the system to collect the nanotubes(see Fig. 4-14).

This process was developed by Dr. Richard Smalley and co-workers at Rice University, who at the time of the discovery of carbon nanotubes, were blasting metals with a laser to produce various metal molecules. When they heard of the existence of nanotubes they replaced the metals with graphite to create multi-walled carbon nanotubes. Later that year the team used a composite of graphite and metal catalyst particles (the best yield was from a cobalt and nickel mixture) to synthesize single-walled carbon nanotubes.

The laser ablation method yields around 70% and produces primarily single-walled carbon nanotubes with a controllable diameter determined by the reaction temperature. However, it is more expensive than either arc discharge or chemical vapor deposition.

In 1996 CNTs were first synthesized using a dual-pulsed laser and achieved yields of $>$ 70wt.% purity (see Fig. 4-15). Samples were prepared by laser vaporization of graphite rods with a 50:50 catalyst mixture of cobalt and nickel at 1,200 ℃ in flowing argon, followed by heat treatment in a vacuum at 1,000 ℃ to remove the C_{60} and other fullerenes. The initial laser vaporization pulse was followed by a second pulse, to vaporize the target more uniformly. The use of two successive laser pulses minimizes the amount of carbon deposited as soot. The second laser pulse breaks up the larger particles ablated by the first one, and feeds them into the growing nanotube structure. The material produced by this method appears as a mat of "ropes", 10-20 nm in diameter and up to 100 μm or more in length. Each rope is found to consist primarily of a bundle of single walled nanotubes, aligned along a common axis. By varying the growth temperature, the catalyst composition, and other process parameters, the average nanotube diameter and size distribution can be varied. Arc-discharge and laser vaporization are currently the principal methods for obtaining small quantities of high quality CNTs. However, both methods suffer from drawbacks. The first is that both methods involve evaporating the carbon source, so it has been unclear how to scale up production to the industrial level using these approaches. The second issue relates to the fact that vaporization methods grow CNTs in highly tangled forms, mixed with unwanted forms of carbon and/or metal species. The CNTs thus produced are difficult to purify, manipulate, and assemble for building nanotube-device architectures for practical applications.

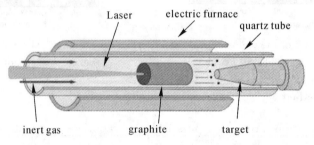

Fig. 4 - 14   Laser system for the synthesis of carbon nanotube

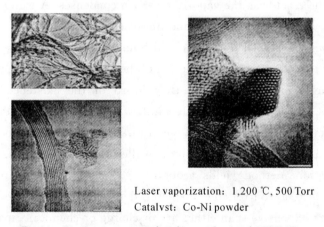

Laser vaporization: 1,200 ℃, 500 Torr
Catalyst: Co-Ni powder

Fig. 4 - 15   Laser system for the synthesis of SWCNTs

4.4.3 Chemical vapor deposition

Chemical vapor deposition of hydrocarbons over a metal catalyst is a classical method that has been used to produce various carbon materials such as carbon fibers and filaments for over twenty years (see Fig. 4 - 16 and Fig. 4 - 17). Large amounts of CNTs or ACNTs (see Fig. 4 - 18) can be formed by catalytic CVD of acetylene over cobalt and iron catalysts supported on silica or zeolite. The carbon deposition activity seems to relate to the cobalt content of the catalyst, whereas the CNTs' selectivity seems to be a function of the pH in catalyst preparation. Fullerenes and bundles of single walled nanotubes were also found among the multi-walled nanotubes produced on the carbon/zeolite catalyst. Some researchers are experimenting with the formation of CNTs from ethylene. Supported catalysts such as iron, cobalt, and nickel, containing either a single metal or a mixture of metals, seem to induce the growth of isolated single walled nanotubes or single walled nanotubes bundles in the ethylene atmosphere. The production of single walled nanotubes, as well as double-walled CNTs, on molybdenum and molybdenum-iron alloy catalysts has also been demonstrated. CVD of carbon within the pores of a thin alumina template with or without a nickel catalyst has been achieved. Ethylene was used with reaction temperatures of 545 ℃ for nickel-catalyzed CVD, and 900 ℃ for an uncatalyzed process. The resultant carbon nanostructures have open ends, with no caps. Methane has also been used as a carbon source. In particular it has been used to obtain "nanotube chips" containing isolated single walled nanotubes at controlled locations. High yields of single-walled nanotubes have been obtained by catalytic decomposition of an H_2/CH_4 mixture over well-dispersed metal particles such as cobalt, nickel, and iron on magnesium oxide at 1,000 ℃. It has been reported that the synthesis of composite powders containing well-dispersed CNTs can be achieved by selective reduction in an H_2/CH_4 atmosphere of oxide solid solutions between a non-reducible oxide such as Al_2O_3 or $MgAl_2O_4$ and one or more transition metal oxides. The reduction produces very small transition metal particles at a temperature of usually >800 ℃. The decomposition of CH_4 over the freshly formed nanoparticles prevents their further growth, and thus results in a very high proportion of single walled nanotubes and fewer multi walled nanotubes.

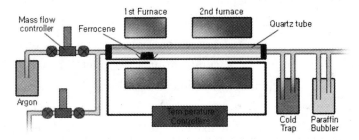

Fig. 4-16   CVD system for the synthesis of carbon nanotube

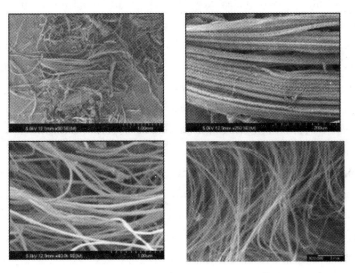

Fig. 4-17   SEM images of MWCNTs using CVD growth

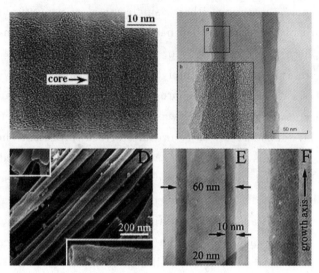

Fig. 4-18 HRTEM images of ACNTs bundles produced by CVD

> **Nanotubes being grown by plasma enhanced chemical vapor deposition.**

The catalytic vapor phase deposition of carbon was first reported in 1959, but it was not until 1993 that carbon nanotubes were formed by this process. In 2007, researchers at the University of Cincinnati (UC) developed a process to grow aligned carbon nanotube arrays of 18 mm length on a First Nano ET3000 carbon nanotube growth system.

During CVD, a substrate is prepared with a layer of metal catalyst particles, most commonly nickel, cobalt, iron, or a combination. The metal nanoparticles can also be produced by other ways, including reduction of oxides or oxides solid solutions. The diameters of the nanotubes that are to be grown are related to the size of the metal particles. This can be controlled by patterned (or masked) deposition of the metal, annealing, or by plasma etching of a metal layer. The substrate is heated to approximately 700 ℃. To initiate the growth of nanotubes, two gases are bled into the reactor: a process gas (such as ammonia, nitrogen or hydrogen) and a carbon-containing gas (such as acetylene, ethylene, ethanol or methane). Nanotubes grow at the sites of the metal catalyst; the carbon-containing gas is broken apart at the surface of the catalyst particle, and the carbon is transported to the edges of the particle, where it forms the nanotubes. This mechanism is still being studied. The catalyst particles can stay at the tips of the growing nanotube during the growth process, or remain at the nanotube base, depending on the adhesion between the catalyst particle and the substrate.

CVD is a common method for the commercial production of carbon nanotubes. For this purpose, the metal nanoparticles are mixed with a catalyst support such as MgO or Al_2O_3 to increase the surface area for higher yield of the catalytic reaction of the carbon feedstock with the metal particles. One issue in this synthesis route is the removal of the catalyst support via an acid treatment, which sometimes could destroy the original structure of the carbon nanotubes. However, alternative catalyst supports that are soluble in water have proven effective for nanotube growth.

If plasma is generated by the application of a strong electric field during the growth process (plasma enhanced chemical vapor deposition), then the nanotube growth will follow the direction of the electric field. By adjusting the geometry of the reactor it is possible to synthesize vertically aligned carbon nanotubes (i.e., perpendicular to the substrate), a morphology that has been of interest to researchers interested in the electron emission from nanotubes. Without the plasma, the resulting nanotubes are often randomly oriented. Under certain reaction conditions, even in the absence of plasma, closely spaced nanotubes will maintain a vertical growth direction resulting in a dense array of tubes resembling a carpet or forest. (see Fig. 4-19)

Of the various means for nanotube synthesis, CVD shows the most promise for industrial-scale deposition, because of its price/unit ratio, and because CVD is capable of growing nanotubes directly on a desired substrate, whereas the nanotubes must be collected in the other growth techniques. The growth sites are controllable by careful deposition of the catalyst. In 2007, a team from Meijo University demonstrated a high-efficiency CVD technique for growing carbon nanotubes from camphor. Researchers at Rice University, until recently led by the late Dr. Richard Smalley, have concentrated upon finding methods to produce large, pure amounts of particular types of nanotubes. Their approach grows long fibers from many small seeds cut from a single nanotube; all of the resulting fibers were found to be of the same diameter as the original nanotube and are expected to be of the same type as the original nanotube. Further characterization of the resulting nanotubes and improvements in yield and length of grown tubes are needed.

CVD growth of multi-walled nanotubes is used by several companies to produce materials on the ton scale, including Nano Lab, Bayer, Arkema, Nanocyl, Nanothinx, Hyperion Catalysis, Mitsui, and Showa Denko.

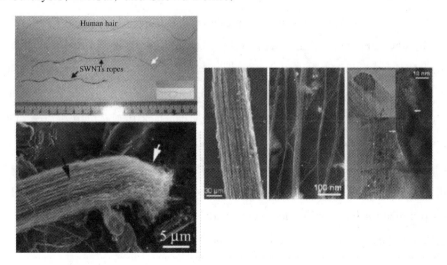

Fig. 4-19 Direct synthesis of SWCNTs strands using CVD

4.4.4 Ball milling

Ball milling and subsequent annealing is a simple method for the production of CNTs. Although it is well established that mechanical attrition of this type can lead to fully nano porous microstructures, it was not until a few years ago that CNTs of carbon and boron nitride were produced from these powders by thermal annealing. Essentially the method consists of placing graphite powder into a stainless steel container along with four hardened steel balls. The container is purged, and argon is introduced. The milling is carried out at room temperature for up to 150 hours. Following milling, the powder is annealed under an inert gas flow at temperatures of 1,400 ℃ for six hours. The mechanism of this process is not known, but it is thought that the ball milling process forms nanotube nuclei, and the annealing process activates nanotube growth. Research has shown that this method produces more multi-walled nanotubes and few single walled nanotubes.

4.4.5 Other methods

CNTs can also be produced by diffusion flame synthesis, electrolysis, use of solar energy, heat treatment of a polymer, and low-temperature solid pyrolysis. In flame synthesis, combustion of a portion of the hydrocarbon gas provides the elevated temperature required, with the remaining fuel conveniently serving as the required hydrocarbon reagent. Hence the flame constitutes an efficient source of both energy and hydrocarbon raw material. Combustion synthesis has been shown to be scalable for high-volume commercial production.

> Natural, incidental, and controlled flame environments.

Fullerenes and carbon nanotubes are not necessarily products of high-tech laboratories; they are commonly formed in such mundane places as ordinary flames, produced by burning methane, ethylene, and benzene, and they have been found in soot from both indoor and outdoor air. However, these naturally occurring varieties can be highly irregular in size and quality because the environment in which they are produced is often highly uncontrolled. Thus, although they can be used in some applications, they can lack in the high degree of uniformity necessary to meet many needs of both research and industry. Recent efforts have focused on producing more uniform carbon nanotubes in controlled flame environments. Nano-C, Inc of Westwood, Massachusetts, is producing flame synthesized single-walled carbon nanotubes. This method has promise for large-scale, low-cost nanotube synthesis, though it must compete with rapidly developing large scale CVD production(see Fig. 4-20).

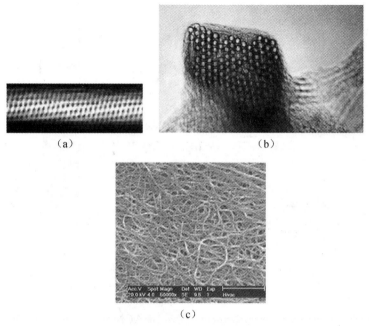

Fig. 4-20 Single-wall carbon nanotube images at different length scales
(a) Scanning tunneling microscope image of a chiral SWNT (image by Clauss W.);
(b) HRTEM image of a nanotube rope (from Thess A. et al., Science 273, 483 (1996), with permission);
(c) Tangled spaghetti of purified SWNT ropes and bundles (Smalley R. E., website)

4.5 The growth mechanism of carbon nanotubes

Since Iijima's landmark paper in 1991, carbon nanotubes (CNTs) have become a hot research topic worldwide. At present, there are three dominant ways to synthesize crystallized CNTs: arc discharge (AD), chemical vapor deposition (CVD) and laser ablation. In recent years, amorphous carbon nanotubes (ACNTs) with diameters in the range of 3 - 60 nm have also been synthesized with CVD and AD methods. The tube-walls of ACNTs are composed of many carbon clusters which are characterized by short-distance order and long-distance disorder. We have modified a traditional arc discharge apparatus with a temperature-controller, creating a system capable of producing both crystallized CNTs and ACNTs through temperature-controlled arc discharge (TC - AD). CNTs formation occurs not only on the core of cathode, but also on the walls of the chamber on a large scale. Using this TC - AD technique we demonstrated a ACNT production rate of about 10 g/h, with tube diameters in the 7 - 20 nm range. In our previous work we have examined the effects of gas, pressure and temperature on the production of ACNTs. According to the literatures, the growth mechanism schematic drawing of CNTs using CVD is shown in Fig. 4 - 21.

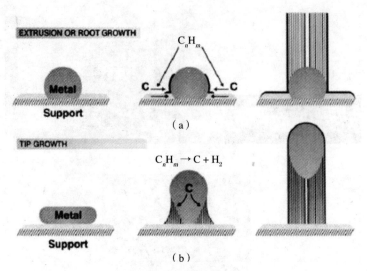

Fig. 4 - 21　The general growth models of CNTs
(a) Bottom growth model;　(b) Top growth model

Here, we only present the growth mechanism of ACNTs(see Fig. 4 - 22).

Based on our characterization of ACNTs synthesized by TC - AD, we propose that the growth mechanism of ACNTs is "open tips and C_n ($n>6$) adding". The growth mechanism of ACNTs proceeds as follows: first, the evaporated carbon atoms form the single-sheet carbon cluster — C_n (namely graphene), generally n is more than several hundred carbon atoms; second, C_n deposits on the catalyst particles to form the base of CNTs; finally, C_n repeatedly deposits on the base of CNTs lead to the growth of CNTs. According to the HRTEM measurement, the length of carbon clusters usually is about 2 - 10 nm. Assuming that the shape of carbon clusters is spherical and composed of single-sheets, the smallest carbon cluster (a diameter of 2 nm) should contain about 220 carbon atoms.

The energy of CNTs consists of structural energy, van der Waals and Coulomb energy. According to the characteristics of ACNTs, the energy of ACNTs should consist of both structural energy and van der Waals.

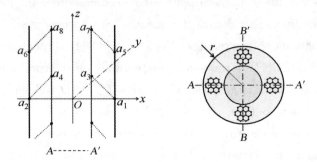

Fig. 4 - 22　Schematics of the physical model of ACNTs

For simplified calculation, we postulate that a carbon cluster is C_{220}, and the angles between carbon cluster and tube-wall are 45° and the distance between each carbon sheet is 0.3 nm. Every crystal unit has 16 carbon atoms and the relative sites of each carbon atom are fixed. There are 8 carbon atoms ($a_1 - a_8$) in plane $A - A'$ and 8 carbon atoms ($a_9 - a_{16}$) in plane $B - B'$. The bond length of C-C is 0.144 nm and we define the radius of ACNTs as r. Therefore, the configuration of every carbon atom can be calculated and they are given by Table 4-1.

Table 4-1 Carbon atoms sites of a crystal unit of ACNT

Sites of carbon atoms	Relative coordinate value
a_1	$(r, 0, 0)$
a_2	$(-r, 0, 0)$
a_3	$(r-1.414, 0, 1.414)$
a_4	$(1.414-r, 0, 1.414)$
a_5	$(r, 0, 0.424)$
a_6	$(-r, 0, 0.424)$
a_7	$(r-1.414, 0, 0.636)$
a_8	$(1.414-r, 0, 0.636)$
a_9	$(0, r, 0)$
a_{10}	$(0, -r, 0)$
a_{11}	$(0, r-1.414, 1.414)$
a_{12}	$(0, 1.414-r, 1.414)$
a_{13}	$(0, r, 0.424)$
a_{14}	$(0, -r, 0.424)$
a_{15}	$(0, r-1.414, 0.636)$
a_{16}	$(0, 1.414-r, 0.636)$

As mentioned previously ACNTs are characterized by short-distance order and long-distance disorder. According to the Tersoff-Brenner potential function, the total potential energy of this system for the short-distance order carbon clusters can be computed. The binding potential energy of carbon clusters among ACNTs can be written as

$$E_b = \sum_i \sum_{j=i+1} [V_R(r_{ij}) - \overline{B}_{ij} V_A(r_{ij})] \qquad (4-1)$$

where E_b is the total potential energy.

i, j are the atom numbers of this system.

r_{ij} is the distance between i and j atoms.

V_R and V_A are the repulsive and attractive pair terms, respectively.

$\overline{B}_{ij}$ is the constant of the binding number, given by

$$\overline{B}_{ij} = (B_{ij} + B_{ji})/2 \qquad (4-2)$$

Here, V_R and V_A are given by

$$V_R(r_{ij}) = f_{ij}(r_i) D_{ij}^{(e)} / (S_{ij} - 1) e^{-\beta_{ij}(r-R(f))\sqrt{2S_{ij}}} \qquad (4-3)$$

and

$$V_A(r_{ij}) = f_{ij}(r_{ij}) D_{ij}^{(e)} S_{ij} / (S_{ij} - 1) e^{-\beta_{ij}(r-R(f))\sqrt{2/S_{ij}}} \qquad (4-4)$$

respectively.

where $D_{ij}^{(e)}$ is the depth of the potential well of the potential function.

$R_{ij}^{(e)}$ is the equilibrium distance between i and j atoms.

β_{ij} and S_{ij} are smoothing parameters, respectively.

$f_{ij}(r)$ is the cut-off function, which one can use to compute the number of the pair potentials and is derivable continuously at the $R_{ij}^{(2)}$. The function $f_{ij}(r)$ restricts the pair potential to nearest neighbors.

Another characteristic of ACNTs is long-distance disorder. The calculation of the interactions among atoms or molecules is carried out by using Lennard-Jones function given by

$$E_{L-J}(r) = 4\varepsilon \left[\left(\frac{\sigma}{r}\right)^{12} - \left(\frac{\sigma}{r}\right)^{6} \right] \qquad (4-5)$$

where $E_{L-J}(r)$ is the nonbinding potential energy between carbon atoms.

ε is the depth of potential well and is 4.204×10^{-3} eV.

σ is length parameter and is 0.337 nm.

While many researchers have applied of the Tersoff-Brenner potential function to CNTs structure, growth mechanisms, hydrogen storage and mechanical capability, the growth mechanism of ACNTs is currently unknown and has not been reported in details. According to the previous investigations of Tersoff-Brenner function in CNTs, we establish a mathematical model for the growth mechanism of ACNTs.

For a single ACNT, the total potential energy should be the sum of the binding potential energy and nonbinding potential energy given by

$$E_{se} = E_b + E_{L-J} \qquad (4-6)$$

where E_{se} is the total potential of ACNT.

E_b is the binding potential energy between inner atoms of carbon clusters.

E_{L-J} is the van der Waals of carbon clusters among ACNTs.

Substituting E_b and E_{L-J}, from Eq.(4-1) and Eq.(4-5) respectively, into Eq.(4-6), a mathematical model for the growth mechanism of ACNTs can be written as:

$$E_{se} = \sum_{i} \sum_{j(>i)} [V_R(r_{ij}) - \overline{B}_{ij} V_A(r_{ij})] + 4\varepsilon \left[\left(\frac{\sigma}{r}\right)^{12} - \left(\frac{\sigma}{r}\right)^{6} \right] \qquad (4-7)$$

Equation (4-7) describes the total potential energy of ACNTs. We think that this model is suitable for these nanotubes which are composed of C-C bonds and van der Waals interactions between graphite sheets or inter-tubes. Nanotubes will grow continuously if the total potential energy reaches a minimum value. This model is incorporated in to a Matlab routine and used to calculate the potential energy of ACNTs of various dimensions at a temperature of 300

K. The 40 ACNTs investigated had diameters ranging from 1 – 25 nm, a length of 100 nm and contained between 200 and 10,000 carbon atoms. The result of these calculations is plotted in Fig. 4 – 23.

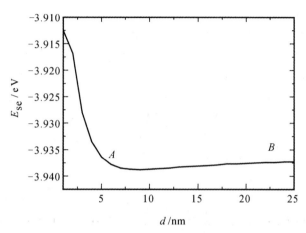

Fig. 4 – 23 Curve of the total energy of ACNT with different diameters

Fig. 4 – 23 shows that the energy of this system decreases rapidly when the diameter of ACNTs is less than 5 nm (before A point), and it indicates that the smaller diameters tubes cannot exist in this system. With increasing tube diameter, the energy of this system increases slowly and the range of ACNTs diameters is about 5 – 25 nm (between A and B points in Fig. 4 – 23). It indicates that this system has low energy and CNTs could exist. When the energy of this system is low and stable, the diameter range of ACNTs is about 5 – 25 nm and is in agreement with experimental diameters range of 7 – 20 nm.

In summary, we propose a growth mechanism of ACNTs, namely, "open tips and $C_n(n>6)$ adding". According to Tersoff-Brenner and Lennard-Jones potential energy functions, a mathematical model for the growth mechanism of ACNTs is established. Model calculations predict theoretical ACNTs diameters in the range of 5 – 25 nm, a result which is in agreement with experimental measurements in the range of 7 – 20 nm.

4.6 Purification methods of carbon nanotube

Purification of CNTs generally refers to the separation of CNTs from other entities, such as carbon nanoparticles, amorphous carbon, residual catalyst, and other unwanted species. The classic chemical techniques for purification have been tried, but they have not been found to be effective in removing the undesirable impurities. Three basic methods have been used with varying degrees of success, namely gas-phase, liquid-phase, and intercalation methods.

Generally, a centrifugal separation is necessary to concentrate the single walled nanotubes in low-yield soot before the micro filtration operation, since the nanoparticles easily contaminate membrane filters. The advantage of this method is that unwanted nanoparticles and amorphous carbon are removed simultaneously and the CNTs are not chemically modified. However 2 – 3 mol nitric acid is useful for chemically removing impurities.

It is now possible to cut CNTs into smaller segments, by extended sonication in concentrated acid mixtures. The resulting CNTs form a colloidal suspension in solvents. They can be deposited on substrates, or further manipulated in solution, and can have many different functional groups attached to the ends and sides of the CNTs.

4.6.1 Gas – phase

The first successful technique for purification of nanotubes was developed by Thomas Ebbesen and coworkers. Following the demonstration that nanotubes could be selectively attached by oxidizing gases these workers realized that nanoparticles, with their defect rich structures might be oxidised more readily than the relatively perfect nanotubes. They found that a significant relative enrichment of nanotubes could be achieved this way, but only at the expense of losing the majority of the original sample.

A new gas-phase method has been developed at the NASA Glenn Research Center to purify gram-scale quantities of single-wall CNTs. This method, a modification of a gas-phase purification technique previously reported by Smalley and others, uses a combination of high-temperature oxidations and repeated extractions with nitric and hydrochloric acid. This improved procedure significantly reduces the amount of impurities such as residual catalyst, and non-nanotube forms of carbon within the CNTs, increasing their stability significantly.

4.6.2 Liquid – phase

The current liquid-phase purification procedure follows certain essential steps:
(1) preliminary filtration to get rid of large graphite particles;
(2) dissolution to remove fullerenes (in organic solvents) and catalyst particles (in concentrated acids);
(3) centrifugal separation;
(4) microfiltration;
(5) chroma to graphy.

It is important to keep the CNTs well-separated in solution, so the CNTs are typically dispersed using a surfactant prior to the last stage of separation.

4.6.3 Intercalation

An alternative approach to purifying multi walled nanotubes was introduced in 1994 by a Japanese research group. This technique made use of the fact that nanoparticles and other graphitic contaminants have relatively "open" structures and can therefore be more readily intercalated with a variety of materials that can close nanotubes. By intercalating with copper chloride, and then reducing this to metallic copper, the research group was able to preferentially oxidize the nanoparticles away, using copper as an oxidation catalyst. Since 1994, this has become a popular method for purification of nanotubes. The first stage is to immerse the crude cathodic deposit in a molten copper chloride and potassium chloride mixture at 400 ℃ and leave it for one week. The product of this treatment, which contains intercalated nanoparticles and graphitic fragments, is then washed in ion exchanged water to remove excess copper chloride and potassium chloride. In order to reduce the intercalated copper chloride-potassium chloride metal, the washed product is slowly heated to 500 ℃ in a mixture of helium and hydrogen and held at this temperature for 1 h. Finally, the material is oxidized in flowing air at a rate of 10 ℃/min to a temperature of 555 ℃. Samples of cathodic soot which have been treated this way consist almost entirely of nanotubes. A disadvantage of this method is that some amount of nanotubes are inevitably lost in the oxidation stage, and the final material may be contaminated with residues of intercalates. A similar purification technique, which involves intercalation with bromine followed by oxidation, has also been described (see Fig.4-24 and Fig.4-25).

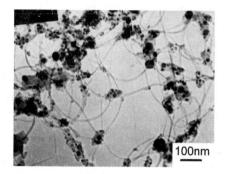
Fig. 4-24　TEM image of as grown SWCNTs

Fig. 4-25　SEM image of purified SWCNTs

4.7　The micro-characterization and properties of carbon nanotube

There are several important modern developments. The atomic force microscope (AFM) and the scanning tunneling microscope (STM) are two early versions of scanning probes that launched nanotechnology. There are other types of scanning probe microscopy, all flowing from the ideas of the scanning confocal microscope developed by Marvin Minsky in 1961 and the scanning acoustic microscope (SAM) developed by Calvin Quate and coworkers in the 1970s, that made it possible to see structures at the nanoscale. The tip of a scanning probe can also be used to manipulate nanostructures (a process called positional assembly). Feature-oriented scanning-positioning methodology suggested by Rostislav Lapshin appears to be a promising way to implement these nanomanipulations in automatic mode. However, this is still a slow process because of low scanning velocity of the microscope. Various techniques of nanolithography such as dip pen nanolithography, electron beam lithography or nanoimprint lithography were also developed. Lithography is a top-down fabrication technique where a bulk material is reduced in size to nanoscale pattern (see Fig. 4-26).

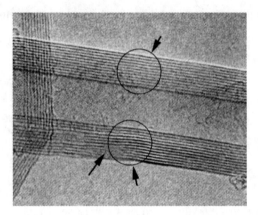

Fig. 4-26　HRTEM image of an MWCNT
The number of graphene layer images on the two edges of the tube differs by 1, implying a scroll structure rather than the assumed-to-be-universal Russian doll morphology
(From Lavin J. G. et al., Carbon 40, 1123-1130 (2002), with permission.)

Another group of nanotechnological techniques include those used for fabrication of nanowires, those used in semiconductor fabrication such as deep ultraviolet lithography, electron beam lithography, focused ion beam machining, nanoimprint lithography, atomic layer deposition, and molecular vapor deposition, and further including molecular self-assembly techniques such as those employing di-block copolymers. However, all of these techniques preceded the nanotech era, and are extensions in the development of scientific advancements rather than techniques which were devised with the sole purpose of creating nanotechnology and which were results of nanotechnology research.

The top-down approach anticipates nanodevices that must be built piece by piece in stages, much as manufactured items are made. Scanning probe microscopy is an important technique both for characterization and synthesis of nanomaterials. Atomic force microscopes and scanning tunneling microscopes can be used to look at surfaces and to move atoms around. By designing different tips for these microscopes, they can be used for carving out structures on surfaces and to help guide self-assembling structures. By using, for example, feature-oriented scanning-positioning approach, atoms can be moved around on a surface with scanning probe microscopy techniques. At present, it is expensive and time-consuming for mass production but very suitable for laboratory experimentation.

In contrast, bottom-up techniques build or grow larger structures atom by atom or molecule by molecule. These techniques include chemical synthesis, self-assembly and positional assembly. Another variation of the bottom-up approach is molecular beam epitaxy or MBE. Researchers at Bell Telephone Laboratories like John R. Arthur, Alfred Y. Cho and Art C. Gossard developed and implemented MBE as a research tool in the late 1960s and 1970s.

Samples made by MBE were key to the discovery of the fractional quantum Hall effect for which the 1998 Nobel Prize in Physics was awarded. MBE allows scientists to lay down atomically-precise layers of atoms and, in the process, build up complex structures. Important for research on semiconductors, MBE is also widely used to make samples and devices for the newly emerging field of spintronics. Newer techniques such as Dual Polarisation Interferometry are enabling scientists to measure quantitatively the molecular interactions that take place at the nano-scale.

However, new therapeutic products, based on responsive nanomaterials, such as the ultra deformable, stress-sensitive transfersome vesicles, are under development and already approved for human use in some countries.

➢ This section tries to give an overview of the many useful and unique properties of CNTs.

1. Electrical conductivity

CNTs can be highly conducting, and hence can be said to be metallic. Their conductivity has been shown to be a function of their chirality, the degree of twist as well as their diameter. CNTs can be either metallic or semi-conducting in their electrical behavior. Conductivity in MWNTs is quite complex. Some types of "armchair"-structured CNTs appear to conduct better than other metallic CNTs. Furthermore, interwall reactions within multi-walled nanotubes have been found to redistribute the current over individual tubes non-uniformly. However, there is no change in current across different parts of metallic single-walled nanotubes. The behavior of the ropes of semi-conducting single walled nanotubes is different, in that the transport current changes abruptly at various positions on the CNTs.

The conductivity and resistivity of ropes of single-walled nanotubes has been measured by placing electrodes at different parts of the CNTs. The resistivity of the single walled nanotubes ropes were of the order of 10^{-4} Ω/cm at 27 ℃. This means that single walled nanotube ropes are the most conductive carbon fibers known. The current density that was possible to achieve was 10^{-7} A/cm^2, however in theory the single-walled nanotube ropes should be able to sustain much higher stable current densities, as high as 10 − 13 A/cm^2. It has been reported that individual single-walled nanotubes may contain defects. Fortuitously, these defects allow the single-walled nanotubes to act as transistors. Likewise, joining CNTs together may form transistor-like devices. A nanotube with a natural junction (where a straight metallic section is joined to a chiral semiconducting section) behaves as a rectifying diode — that is, a half-transistor in a single molecule. It has also recently been reported that single walled nanotubes can route electrical signals at speeds up to 10 GHz when used as interconnects on semi-conducting devices.

2. Strength and elasticity

The carbon atoms of a single sheet of graphite form a planar honeycomb lattice, in which each atom is connected via a strong chemical bond to three neighboring atoms. Because of these strong bonds, the basal plane elastic modulus of graphite is one of the largest of any known material. For this reason, CNTs are expected to be the ultimate high-strength fibers. Single-walled nanotubes are stiffer than steel, and are very resistant to damage from physical forces. Pressing on the tip of a nanotube will cause it to bend, but without damage to the tip. When the force is removed, the nanotube returns to its original state. This property makes CNTs very useful as probe tips for very high-resolution scanning probe microscopy. Quantifying these effects has been rather difficult, and an exact numerical value has not been agreed upon.

Using atomic force microscopy, the unanchored ends of a freestanding nanotube can be pushed out of their equilibrium position, and the force required to push the nanotube can be measured. The current Young's modulus value of single-walled nanotubes is about 1 TPa, but this value has been widely disputed, and a value as high as 1.8 TPa has been reported. Other values significantly higher than that have also been reported. The differences probably arise through different experimental measurement techniques. Others have shown theoretically that the Young's modulus depends on the size and chirality of the single-walled nanotubes, ranging from 1.22 - 1.26 TPa. They have calculated a value of 1.09 TPa for a generic nanotube. However, when working with different multi walled nanotubes, others have noted that the modulus measurements of multi-walled nanotubes using AFM techniques do not strongly depend on the diameter. Instead, they argue that the modulus of the multi walled nanotubes correlates to the amount of disorder in the nanotube walls. Not surprisingly, when multi-walled nanotubes break, the outermost layers break first.

3. Thermal conductivity and expansion

CNTs have been shown to exhibit superconductivity below 20 K (aaprox. $-$ 253 ℃). Research suggests that these exotic strands, already heralded for their unparalleled strength and unique ability to adopt the electrical properties of either semiconductors or perfect metals, may someday also find applications as miniature heat conduits in a host of devices and materials. The strong in-plane graphitic carbon - carbon bonds make them exceptionally strong and stiff against axial strains. The almost zero inter-plane thermal expansion but large inter-plane expansion of single walled nanotubes implies strong inter-plane coupling and high flexibility against non-axial strains.

Many applications of CNTs, such as in nanoscale molecular electronics, sensing and actuating devices, or as reinforcing additive fibers in functional composite materials, have been proposed. Reports of several recent experiments on the preparation and mechanical characterization of CNT-polymer composites have also appeared. These measurements suggest modest enhancements in strength characteristics of CNT-embedded matrixes as compared to bare polymer matrixes. Preliminary experiments and simulation studies on the thermal properties of CNTs show very high thermal conductivity. It is expected, therefore, that nanotube reinforcements in polymeric materials may also significantly improve the thermal and thermomechanical properties of the composites.

4. Field emission

Field emission results from the tunneling of electrons from a metal tip into vacuum, under application of a strong electric field. The small diameter and high aspect ratio of CNTs is very favorable for field emission. Even for moderate voltages, a strong electric field develops at the free end of supported CNTs because of their sharpness. This was observed by de Heer and co-workers at EPFL in 1995. He also immediately realized that these field emitters must be superior to conventional electron sources and might find their way into all kind of applications, most importantly flat-panel displays. It is remarkable that after only five years Samsung actually realized a very bright color display, which will be shortly commercialized using this technology. Studying the field emission properties of multi-walled nanotubes, Bonard and co-workers at EPFL observed that together with electrons, light is emitted as well. This luminescence is induced by the electron field emission, since it is not detected without applied potential. This light emission occurs in the visible part of the spectrum, and can sometimes be seen with the naked eye.

5. High aspect ratio

CNTs represent a very small, high aspect ratio conductive additive for plastics of all types. Their high aspect ratio means that a lower loading of CNTs is needed compared to other conductive additives to achieve the same electrical conductivity. This low loading preserves more of the polymer resins' toughness, especially at low temperatures, as well as maintaining other key performance properties of the matrix resin. CNTs have proven to be an excellent additive to impart electrical conductivity in plastics. Their high aspect ratio, about 1,000 : 1 imparts electrical conductivity at lower loadings, compared to conventional additive materials such as carbon black, chopped carbon fiber or stainless steel fiber.

6. Highly absorbent

The large surface area and high absorbency of CNTs make them ideal candidates for use in air, gas, and water filtration. A lot of research is being done in replacing activated charcoal with CNTs in certain ultra high purity applications.

Activities and problems for students

Activities.

(1) Was carbon nanotube discovered in 1991?
(2) How many different kinds of carbon nanotubes are there?
(3) How to prepare carbon nanotube in large scale?
(4) How many synthesis methods of carbon nanotube are there?

Problems.

What's the unique properties of carbon nanotubes?

Chapter 5　The applications of carbon nanotubes

5.1　Overview of potential and current applications

　　The special nature of carbon combined with the molecular perfection of single-walled carbon nanotubes (SWCNTs) to endow them with exceptional material properties, such as very high electrical and thermal conductivity, strength, stiffness, and toughness. No other element in the periodic table bonds to itself in an extended network with the strength of the carbon-carbon bond. The delocalized pi-electron donated by each atom is free to move about the entire structure, rather than remain with its donor atom, giving rise to the first known molecule with metallic-type electrical conductivity. Furthermore, the high-frequency carbon-carbon bond vibrations provide an intrinsic thermal conductivity higher than even that of diamond. In most conventional materials, however, the actual observed material properties such as strength, electrical conductivity, etc. are degraded very substantially by the occurrence of defects in their structures. For example, high-strength steel typically fails at only about 1% of its theoretical breaking strength. Carbon nanotubes (CNTs), however, achieve values very close to their theoretical limits because of their molecular perfection of structure.

　　This aspect is part of the unique story of CNTs. CNT is an example of true nanotechnology: they are under 100 nm in diameter, but are molecules that can be manipulated chemically and physically in very useful ways. They open an incredible range of applications in materials science, electronics, chemical processing, energy conversion, and many other fields. CNTs have extraordinary electrical conductivity, heat conductivity, and mechanical properties. They are probably the best electron field-emitter. They are polymers of pure carbon and can be reacted and manipulated using the well-known, tremendously rich chemistry of carbon. This provides opportunity to modify their structure, and to optimize their solubility and dispersion. Very significantly, CNTs are molecularly perfect, which means that they are normally free of property-degrading flaws in the nanotube structure (see Fig. 5 - 1). Their material properties can therefore approach close to the very high levels intrinsic to them. These extraordinary characteristics give CNTs potential in numerous applications.

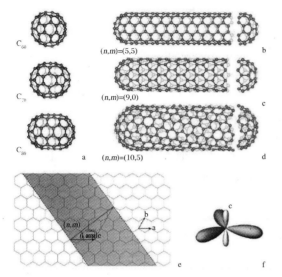

Fig. 5-1 Schematic drawing of the structure of CNTs

Fig. 5-2 Cover of *American Scientist* in 1997 (space elevators)

The strength and flexibility of CNTs makes them of potential use in controlling other nanoscale structures, which suggests they will have an important role in nanotechnology engineering. The highest tensile strength of an individual multi-walled CNT (MWCNT) has been tested to be 63 GPa. CNTs were found in Damascus steel, possibly helping to account for the legendary strength of the (arguably the most ancient) swords made of it. Because of CNT's superior mechanical properties, many structures have been proposed ranging from everyday items like clothes and sports gear to combat jackets and space elevators (see Fig. 5-2). However, the space elevator will require further efforts in refining CNT technology, as the practical tensile strength of CNTs can still be greatly improved.

For perspective, outstanding breakthroughs have already been made. Pioneering work led by Ray H. Baughman at the NanoTech Institute has shown that SWCNTs and MWCNTs can produce materials with toughness unmatched in the man-made and natural world.

1. In electrical circuits

CNTs have many properties — from their unique dimensions to an unusual currentconduction mechanism — that make them ideal components of electrical circuits. For example, they have shown to exhibit strong electron-phonon resonances, which indicate that under certain direct current (DC) bias and doping conditions their current and the average electron velocity, as well as the electron concentration on the tube oscillate at terahertz frequencies. These resonances could potentially be used to make terahertz sources or sensors. CNTs based transistors, also known as CNT-(field-effect transistor) FETs, have been made that operate at room temperature and that are capable of enabling digital switching using a single electron. Historically, one major obstacle to realization of CNTs has been the lack of technology for mass production. However, in 2001 IBM researchers demonstrated how nanotube transistors can be grown in bulk, somewhat like silicon transistors. Their process is called "constructive destruction" which includes the automatic destruction of defective nanotubes on the wafer. The IBM process has been developed further and single-chip wafers with over ten billion correctly aligned CNT junctions have been created. In addition it has been demonstrated that incorrectly aligned CNTs can be removed automatically using standard photolithography equipment.

The first nanotube integrated memory circuit was made in 2004. One of the main challenges has been regulating the conductivity of CNTs. Depending on subtle surface features a CNT may act as a plainconductor or as a semiconductor. A fully automated method has however been developed to remove non-semiconductor tubes. Recently, collaborators in US and China at Duke University and Peking University announced a new CVD recipe involving a combination of ethanol and methanol gases and quartz substrates resulting in horizontally aligned arrays of 95%-98% semiconducting nanotubes. This is considered a large step towards the ultimate goal of producing perfectly aligned 100% semiconducting CNTs for mass production of electronic devices.

Another method to produce CNT transistors was to use random networks of them. By doing so one averages all of their electrical differences and one can produce devices in large scale (wafer level). This approach was first patented by Nanomix Inc. in June 2002. It was first published in the academic literature by the United States Naval Research Laboratory in 2003 through independent research work. This approach also enabled Nanomix to make the first transistor on a flexible and transparent substrate.

CNTs are usually grown on nanoparticles of magnetic metal (Fe, Co), which facilitates production of electronic (spintronic) devices. In particular control of current through a FET by magnetic field has been demonstrated in such a single-tube nanostructure. Large structures of CNTs can be used for thermal management of electronic circuits. An approximately 1 mm thick CNTs layer was used as a special material to fabricate coolers, this materials has very low density, 20 times lower weight than a similar copper structure, while the cooling properties are similar for the two materials.

2. Secondary batteries

A paper battery is a battery engineered to use a paper-thin sheet of cellulose infused with aligned CNTs. The nanotubes act as electrodes, allowing the storage devices to conduct electricity. The battery, which functions as both a lithium-ion battery and a supercapacitor, can provide a long and steady power output comparable to a conventional battery, as well as a supercapacitor's quick burst of high energy. And the conventional battery contains a number of separate components, but the paper battery integrates all of the battery components in a single structure, making it more energy efficient.

3. Solar cells

One of the promising applications of SWCNTs is their use in solar panels, due to their strong UV/Vis-NIR absorption characteristics. Researchers have shown that they can provide a sizeable increase in efficiency, even at their current unoptimized state. Solar cells developed at the New Jersey Institute of Technology use a CNT complex, formed by a mixture of CNTs and carbon buckyballs (known as fullerenes or C_{60}) to form snake-like structures. Buckyballs (or C_{60}) trap electrons, although they can't make electrons flow. Add sunlight to excite the polymers, and the buckyballs will grab the electrons. Nanotubes, behaving like copper wires, will then be able to make the electrons or current flow. Additional research has been conducted on creating SWCNTs hybrid solar panels to increase the efficiency further. These hybrids are created by combining SWCNTs with photoexcitable electron donors to increase the number of electrons generated. It has been found that the interaction between the photoexcited porphrin and SWCNTs generates electro-hole pairs at the SWCNT surfaces. This phenomenon has been observed experimentally, and contributes practically to an increase in efficiency up to 8.5%.

4. Ultracapacitors

MIT laboratory for electromagnetic and electronic systems uses nanotubes to improve ultracapacitors. The activated charcoal used in conventional ultracapacitors has many small hollow spaces of various size, which create together a large surface to store electric charges. But as charge is quantized into elementary charges, i.e. electrons, and each such elementary charge needs a minimum space, a significant fraction of the electrode surface is not available for storage because the hollow spaces are not compatible with the charge's requirements. With a nanotube electrode the spaces may be tailored to size — few too large or too small — and consequently the capacity should be increased considerably.

5. Radar wave absorption

Radars work in the microwave frequency range, which can be absorbed by CNTs, especially MWCNTs and amorphous CNTs (ACNTs). Applying the MWCNTs to the aircraft would cause the radarwave to be absorbed and therefore seem to have a smaller signature. One such application could be to paint the nanotubes onto the plane. Recently there has been some works done at the University of Michigan regarding CNTs usefulness as stealth technology on aircraft. It has been found that in addition to the radar absorbing properties, the nanotubes neither reflect nor scatter visible light, making it essentially invisible at night, much like painting current stealth aircraft black except much more effective. Current limitations in manufacturing, however, mean that current production of nanotube-coated aircraft is not possible. One theory to overcome these current limitations is to cover small particles with the nanotubes and suspend the nanotube-covered particles in a medium such as paint, which can then be applied to a surface, like a stealth aircraft.

6. Medical applications

The nanotube's versatile structure allows it to be used for a variety of tasks in and around the body. Although often seen especially in cancer-related incidents, CNTs are often used as a vessel for transporting drugs into the body. The nanotube application potentially allows for the drug dosage to be lowered by localizing its distribution. The nanotube commonly carries the drug in one of two ways: the drug can be attached to the side or trailed behind, or the drug can actually be placed inside the nanotube. Both of these methods are effective for the delivery and distribution of drugs inside the body.

7. Other applications

CNTs have been implemented in nanoelectromechanical systems, including mechanical memory elements and nanoscale electric motors. CNTs have been proposed as a possible gene delivery vehicle and for use in combination with radiofrequency fields to destroy cancer cells.

In 2005, Nanomix Inc. placed on the market a hydrogen sensor which integrated CNTs on a silicon platform. Since then Nanomix has been patenting many such sensor applications in the field of carbon dioxide, nitrous oxide, glucose, DNA detection, etc. Eikos Inc. of Franklin, Massachusetts, and Unidym Inc. of Silicon Valley, California are developing transparent, electrically conductive films of CNTs to replace indium tin oxide (ITO). CNT films are substantially more mechanically robust than ITO films, making them ideal for high-reliability touchscreens and flexible displays. Printable water-based inks of CNTs are desired to enable the production of these films to replace ITO. Nanotube films show promise for use in displays for computers, cell phones, PDAs and ATMs. A nanoradio, a radio receiver consisting of a single nanotube, was demonstrated in 2007. In 2008 it was shown that a sheet of nanotubes can operate as a loudspeaker if an alternating current is applied. The sound is not produced through vibration but thermoacoustics.

CNTs have been made in weaving them into clothes to create stab-proof and bulletproof clothing. The nanotubes would effectively stop the bullet from penetrating the body, although the bullet's kinetic energy would likely cause broken bones and internal bleeding. A flywheel made of CNTs could be spun at extremely high speed on a floating magnetic axis, and potentially store energy at a density approaching that of conventional fossil fuels. Since energy can be added to and removed from flywheels very efficiently in the form of electricity, this might offer a way of storing electricity, making the electrical grid more efficient and various kinds of power suppliers (like wind turbines) more useful in meeting energy needs. Nitrogen-doped CNTs may replace platinum catalysts used to reduce oxygen in fuel cells. A forest of vertically-aligned nanotubes can reduce oxygen in alkaline solution more effectively than platinum, which has been used in such applications since the 1960s. The nanotubes have the added benefit of not being subject to carbon monoxide poisoning[25].

5.2 Functionalized carbon nanotubes

5.2.1 Functionalization and dispersion of CNTs

The first major commercial application of MWCNTs is their use as electrically conducting components in polymer composites. Depending on the polymer matrix, conductivities of $0.01 - 0.1$ S·cm^{-1} can be obtained for 5% loading, much lower conductivity levels suffice for dissipating electrostatic charge. The low loading levels and the nanofiber morphology of the MWCNTs allow electronic conductivity to be achieved while avoiding or minimizing degradation of other performance aspects, such as mechanical properties and the low melt flow viscosity needed for thin-wall molding applications. Cost dictates the use of MWCNTs rather than SWCNTs, but unbundled SWCNTs should enable lower percolation levels, reducing the required loading levels further. A percolation threshold of 0.1%–0.2% has been reported for SWCNTs in epoxy, one-tenth that of commercially available 200-nm-diameter vapor-grown carbon fibers. The shielding of electromagnetic radiation from cell phones and computers by using molded SWCNT and MWCNT composites is also a potentially lucrative application. The critical challenges lie in uniformly dispersing the nanotubes, achieving nanotube-matrix adhesion that provides effective stress transfer, and avoiding intratube sliding between concentric tubes within MWCNTs and intrabundle sliding within SWCNT ropes.

Before using it in any fields, the functionalization and/or dispersion of CNTs is needed to do. Below are the functionalization and dispersion processes of CNTs in our experiments.

Carboxylic MWCNTs (f-MWCNTs) were prepared using a mixed solution (HNO_3 : H_2SO_4 = 1 : 3, V/V) by ultrasonication for 5 h at 60 ℃. They were then rinsed with deionized water until the solution was neutral. Finally, they were baked in a vacuum oven at 100 ℃ for 8 h.

Preparation of ionic liquid (IL): first, pyridine (40.0 g) was added into a 250 mL round-bottomed flask that was sealed with turned-mouth rubber stopper. Then nitrogen was pumped and filled into the flask three times repeatedly. At room temperature and under magnetic stirring, 75.4 g of butane bromide was added into the flask using micro-syringe, and then heated to 50 ℃ using water-bath apparatus and stirred for 12 h. The product was added to 200 mL of ethyl acetate, the process of oscillation and washing operation were carried out five times. And then this product was put into 200 mL of acetonitrile for recrystallization. Finally, the product was dried in a vacuum oven at 80 ℃ for 12 h.

5.2.2 Preparation of the functionalized-CNT composites

Authors have carried out a series of synthesis processes of CNTs based composites, such as carboxymethyl cellulose (CMC)/MWCNT, chitosan (CS)/MWCNT and Fe_3O_4/MWCNT composites. The detail parameters of these synthesis processes are as follows.

Preparation of CMC/MWCNT composite: first, 1 mg of MWCNTs and 20 g of IL were added into a 100 mL round-bottomed flask under ultrasonication for 1 h at 80 ℃. Then 20 g of IL containing 5 wt.% CMC was added to the above solution, and the mixture solution was sonicated for 1 h, followed by adding 20 mL of deionized water. The reaction solution was filtrated with 0.22 μm mixed-fiber membranes. The resulting black product was dispersed in deionized water again after leaching and this process was repeated five times. Finally, the obtained CMC/MWCNT composites were distributed into 5 mL of deionized water.

Preparation of CS/MWCNT composite: CS powders were sonicated and stirred in a 0.1 M acetic acid solution containing 1% tween 80 and a 20% Na_2SO_4 solution. This solution was added in 5 mL/min increments over a period of 1 h. CS nanoparticles were separated by centrifugation at 3,000 rpm and dried at room temperature. 1 mg·mL^{-1} CS solution containing 1% acetic acid solution was sonicated for 30 min, and mixed with an equal volume of 1 mg·mL^{-1} c-MWCNTs solution. The blended solution was stirred for 4 h, and washed with 1% acetic acid solution to remove unreacted CS nanoparticles. After baking in the oven at 40 ℃ for 12 h, MWCNTs/CS composite were obtained by electrostatic self-assembly. 1 mg·mL^{-1} MWCNTs/CS ethanol solution was deposited onto the clean surface of a GC electrode using a micro syringe. The electrode was polished with 0.3 μm and 0.05 μm alumina slurry and rinsed with deionized water, sonicated with acetone, ethanol and deionized water for 4 min each, and finally dried with an infrared lamp to obtain a finished MWCNTs/CS/GC electrode.

Preparation of MWCNT/Fe_3O_4 composite: 0.06 g acid-treated MWCNTs and 0.46 g Fe_3O_4 nanoparticles were dispersed in 100 mL deionized water and then the above solution contained 0.02 mol·L^{-1} Fe_3O_4 was sonicated for 4 h at 80 ℃. Then, the solution was dried under vacuum condition at 60 ℃ for 6 h. Finally, MWCNT/Fe_3O_4 composite (CFe_3O_4 = 0.02 mol·L^{-1}) was obtained. According to the above steps, MWCNT/Fe_3O_4 composite were prepared using 0.07 mol·L^{-1} and 0.1 mol·L^{-1} Fe_3O_4 nanoparticles (the content of MWCNTs is the same value).

5.3 Applications of carbon nanotubes in energy conversion

Research has shown that CNTs have the highest reversible capacity of any carbon material for use in lithium ion batteries. In addition, CNTs are outstanding materials for super capacitor electrodes and are now being marketed for this application. CNTs also have applications in a variety of fuel cell components. They have a number of properties, including high surface area and thermal conductivity, which make them useful as electrode catalyst supports in PEM fuel cells. Because of their high electrical conductivity, they may also be used in gas diffusion layers, as well as current collectors. CNTs' high strength and toughness-to-weight characteristics may also prove valuable as part of composite components in fuel cells that are deployed in transport applications, where durability is extremely important.

5.3.1 Hydrogen storage

Hydrogen is a clean, convenient, versatile fuel source that easily converts to a desired form of energy without releasing harmful emissions. Hydrogen energy and its applications have attracted many researchers to study. However, hydrogen storage and transportation still is a barrier to be overcome. The US DOE has established a hydrogen storage target of 9 wt.% for the year 2015. At present, three classes of materials are being investigated for hydrogen storage: ① metal hydrides (e.g. $LaNi_5$, Mg_2Ni and $NaAl$) reversible solid-state materials regenerated on-board. ② Chemical hydrides (e.g. chemical hydrides slurry and solution), hydrogen is released via chemical reaction (usually with water); byproduct is regenerated off-board. ③ Carbon-based materials (e.g. CNTs, grapheme etc.) and fibers and super-high surface area activated carbon) reversible solid-state materials regenerated on-board. Hydrogen storage technologies can be divided into physical storage, where hydrogen molecules are stored (including pure hydrogen storage via compression and liquefication), and chemical storage, where hydrides are stored (such as metal hydrides).

CNTs have the intrinsic characteristics desired in material used as electrodes in batteries and capacitors, two technologies of rapidly increasing importance. CNTs have a tremendously high surface area, good electrical conductivity, and very importantly, their linear geometry makes their surface highly accessible to the electrolyte.

Targets were set by the Freedom CAR Partnership in 2002 between the United States Council for Automotive Research (USCAR) and U.S. DOE (targets assume a 5 kg H_2 storage system). The 2005 targets were not reached. It is important to note that these targets are for the hydrogen storage system, not the hydrogen storage material. Thus while a material may store 6 wt.% H_2, a working system using that material may only achieve 3 wt.% when the weight of tanks, temperature and pressure control equipment, etc., is considered. System densities are often around half those of the working material.

For studying the hydrogen storage performance of CNTs, especially ACNTs, a series experiments have been carried out in the author's group.

> ACNTs hydrogen storage.

Recently, many researchers have reported how to store hydrogen easily and cheaply, using CNTs and nanofibers. Many experimental investigations of hydrogen uptake in CNTs have been carried out. In 1997, Dillon et al. first reported the hydrogen adsorption capacity of as-prepared SWCNTs is about 5 wt.% - 10 wt.%. Dillon et al. showed that the short SWCNTs with open ends can adsorb 3.5 wt.% - 4.5 wt.% hydrogen under ambient conditions in several minutes. Ye et al. reported that a ratio of H to C atoms of about 1.0 was obtained for crystalline ropes of SWCNTs at 80K and pressure >12 MPa. Liu et al. showed that SWCNTs can adsorbed hydrogen about 4.2 wt.% under pressure 10 MPa and room temperature, but this result cannot be reproduced according to reference. Chembers et al. reported that tubular, platelet, and herringbone forms of graphite nanofibers (GNF) can adsorbed 11 wt.%, 45 wt.% and 67 wt.% hydrogen, respectively, at room temperature under a pressure of about 12 MPa. Chen et al. reported that their Temperature-Programmed Desorption (TPD) experiments show a high hydrogen uptake of 20 wt.% and 14 wt.% in milligram quantities of Li-doped and K-doped MWCNTs, respectively, under ambient pressure and moderate temperature (200 – 400 ℃). Bachmatiuk et al. reported 2.2 wt.% of H_2 uptake in B-doped SWCNTs. Similarly, Pt-dispersed SWCNTs is reported to have a storage capacity of 3.03 wt.%. Ci et al. reported that ACNTs after annealing at 1,700 – 2,200 ℃ can adsorb H_2 up to 3.98 wt.% at room temperature under modest pressure (10 MPa).

5.3.2 Lithium ion secondary batteries

Lithium ion rechargeable battery is an attractive power source used in mobile electronic equipment, and electric vehicles because of its high energy density, superior cycle life and environmental friendship. This battery incorporates lithiated metal oxides (typically $LiCoO_2$, $LiMn_2O_4$ or $LiFePO_4$) as the cathode materials and graphite as the anode materials. Graphite (a theoretical capacity of 372 mA·h·g^{-1}) could not satisfy further need for higher energy density battery. At present, various types of carbon have been investigated extensively among which CNTs have attracted much attention because of their unique structure and high capacity. It was found that the capacity of CNTs is strongly dependent on the morphology, structure and the methods of synthesis. Generally, the capacity of CNTs is about 600 mA·h·g^{-1}. To enhance the capacity and cycle life, some improvement treatments including ball-milling, doping, etching or oxidation have been adopted and the discharge capacity increased over 1,000 mA·h·g^{-1}. However, the reversible capacity is still low and the degradation is large after several cycles. Therefore, seeking a large reversible capacity and long cycle life Li-ion battery materials is still a pursuing task. During recent years, we have studied the synthesis and characteristics of ACNTs by a modified arc discharge and the application of ACNTs on lithium ion batteries. Here, the application of ACNTs as the anode materials for lithium ion batteries and the intercalation mechanism were deeply investigated.

A modified arc discharging furnace was used for the synthesis of ACNTs. Compared to the traditional electric arc method, this furnace can control the temperature of reactive chamber from 25 – 900 ℃. Co-Ni (1 : 1 in weight ratio) alloy powders were used as catalyst and H_2 was selected as a buffer gas. 10 g of ACNTs per hour can be obtained with this equipment.

5.4 Application of carbon nanotubes in military

5.4.1 Electromagnetic wave absorber

The pollution of electromagnetic (EM) radiation in the frequency range of 1 – 20 GHz has been a serious concern in modern society. Hence, more and more economical EM wave absorbers with wide absorption bandwidth and effective absorption properties have been strongly demanded. It is believed that the absorption properties associate greatly with the structure (such as single- and multi-layer structures) of the EM wave absorbers and were determined by the complex permittivity ($\varepsilon_r = \varepsilon' - j\varepsilon''$) and permeability ($\mu_r = \mu' - j\mu''$), where ε_r is the relative complex permittivity and μ_r is the relative complex permeability. In recent years, the rare metal elements (RMs) and MWCNTs have been extensively used as the EM wave absorbers. Sun and coworkers investigated the EM wave absorbing properties of MWCNTs modified by RM oxides, and they found that the absorbing properties were remarkably improved. Zheng et al. reported that the magnetic, dielectric loss and reflection loss of Co/MWCNTs composites are improved compared to that of pure MWCNTs or pure cobalt powder.

For expanding the novel application in military field, the composites based on modified CNTs (REs modification) were synthesized by using resin such as polyvinyl chloride (PVC).

5.4.2 Electromagnetic shielding structure

Recently, on the basis of the superior electronic and mechanical properties of CNTs, the CNTs have been the focus of considerable research and development for uses in nanoscale electronic and optoelectronic applications, such as integrated circuit (IC) interconnections, energy storage, nanosensor, and shielding material.

For the application of being shielding material packaging in the optical transceiver modules, electromagnetic interference (EMI) is one of the major concerns to maintain good signal quality of over gigabit transmission rate. Designing a high electromagnetic shielding package is desirable to improve the EMI performance of the optical transceiver modules. It is well known that metallic package provides an excellent shielding effectiveness (SE). However, the characteristics of low cost and eased manufacturing have promoted the plastic composite package as the most suitable material for fabricating the optical transceiver modules for using in the fiber to the home (FTTH) applications.

But plastics are inherently transparent to electromagnetic (EM) radiation and provide no shielding effect against radiation emissions. In order to improve the EM shielding for the plastic packaging, MWCNTs were to be mixed into the plastic hosts to get an adequate EM shielding property to be the conductive fillers in this study. The reasons of choosing MWCNTs to be the fillers were their known properties of high electrical conductivity, nanoscale diameter, high aspect ratio, and possibly strengthened mechanical properties.

However, the high aspect ratio and low ionic character of MWCNTs are not easily dispersed within the plastic hosts. The lack of dispersing ability in the polymer matrices is caused by internal van der Waals force among the MWCNTs and their consequent aggregation. Without a fine dispersion, the MWCNTs may form local clusters and poor homogeneity existed in the MWCNT composite. As a consequence, the weight percentage of the added MWCNTs is required in order to achieve a good electrical conductivity and a comparable SE. For developing a cost effective material, a fine dispersion of MWCNTs in the polymer matrices and association with dispersion mechanism is essential.

Electrically conducting polymer composites have received much attention recently compared to conventional metal-based EMI shielding materials, because of their light weight, resistance to corrosion, flexibility and processing advantages. In the reference, researchers found that SWCNT — epoxy composites had excellent EMI shielding performance in the frequency range of 10 MHz to 1.5 GHz. However, shielding in the range of 8.2 - 12.4 GHz (the so-called X-band) is more important for many military and commercial applications. For example, Doppler, weather radar, TV picture transmission, and telephone microwave relay systems lie in X-band.

The EMI shielding effectiveness of a composite material depends mainly on the filler's intrinsic conductivity, dielectric constant, and aspect ratio. The small diameter, high aspect ratio, high conductivity and mechanical strength of CNTs make them an excellent option to create conductive composites for high-performance EMI shielding materials at very low filling. Despite several studies on the EMI of MWCNT composites, EMI studies for SWCNT composites are few. The electrical properties of small diameter SWCNTs are distinctly different from their larger diameter MWCNT counterparts. Small diameter (i.e., $1<d<2$ nm) SWCNTs can be either metallic or semiconducting depending on their chirality integers (n, m). In addition, small diameter metallic SWCNTs have been found to be exceptional metals, even exhibiting ballistic transport at low temperature. On the other hand, MWCNTs, because of the larger inherent diameter of SWCNTs present in the concentric tube shells (i.e., $d>5$ nm), should be zero gap semiconductors or exhibit very weak band overlap leading to weak semi-metallic behavior. Thus, per unit wt.% added to the polymer host, the nature of the EMI shielding properties of MWCNTs and SWCNT-polymer composites are expected to be altogether different.

5.5 Application of carbon nanotubes in electrochemistry

Since S. Iijima's landmark paper in 1991, CNTs have attracted scientists' extensive research attention due to their remarkable physical and chemical properties, and the potential application in the fields of electronic devices, catalysts and biomedicine. In recent years, the studies of CNTs and biological molecules composite for the purpose of biological applications are rapidly increasing. It has become a new and hot research topic owing to integrating the performances both biologically active molecules (enzymes, proteins, DNA, etc.) and CNTs. Cellulose, as an inexhaustible, natural and renewable polymer material resource, grows and exists in green plants. In addition, cellulose possesses the merits of low cost, rich resource and good biocompatibility. Therefore, cellulose/MWCNT composite has good potential application in the fields of biomaterials and electrochemistry.

Ionic liquid (IL), as a new type of material, is considered to be an environmentally friendly green solvent. IL has many advantages such as high thermostability, high ionic conductivity, low toxicity, non-inflammability, non-volatility and non-explosive ability. It has the nature of both liquid and ionic solvent compared to the solid and conventional liquid material. And it also has lower melting point (not higher than 100 ℃) than the molten salt. Recently, Rogers et al. reported that the IL of bromide 1-butyl-3-methylimidazolium ([BMIM]Br) which destructed the hydrogen bonds between cellulose molecules could directly dissolve cellulose without any processing. Therefore, these nano-composites prepared using CNTs and the cellulose dissolved with ionic liquid are expected to have widely potential applications.

Cellulose/MWCNT composite as a new type of material has unique structure and excellent physical and chemical properties. It combines the excellent performances of both cellulose (good biocompatibility) and CNTs (high surface area and strength), and also it is a novel functional and structural material. Cellulose/MWCNT composites have broad application prospect in the fields of biosensors and electrochemistry. IL can directly dissolve cellulose and cannot affect the cooperative properties of the MWCNTs and cellulose. It is significant to reduce the costs and process steps of these composites. Carboxymethyl cellulose sodium (CMC) is a sodium salt of cellulose. As far as we know, the study of CMC/MWCNT biocomposites has not been previously reported.

In this section, a biocompatible CMC/MWCNT biocomposite was synthesized using MWCNTs functionalized with CMC dissolved in IL. Especially, the electrocatalytic performance of the glass carbon electrode coated with CMC/MWCNT biocomposite towards H_2O_2 was measured. The effects of CNT content and ultrasonic time on the electrochemical performance were also studied.

The pristine MWCNTs used (average diameter: 11 nm, purity: ≥95%) came from Beijing Nano Technology Co., Ltd. (China). CMC was purchased from Tianjin Dongli Chemical Reagent Co., Ltd. (China). Concentrated sulfuric acid, concentrated nitric acid, pyridine, n-butane bromide and acetonitrile/ethyl acetate were supplied by National Medicine Inc. of Shaanxi Province. All the reagents were commercially available and of analytical grade. The chemicals were all used as received without further purification.

Carboxylic MWCNTs (f-MWCNTs) were prepared using a mixed solution (HNO_3 : H_2SO_4 = 1 : 3, V/V) by ultrasonication for 5 h at 60 ℃. They were then rinsed with deionized water until the solution was neutral. Finally, they were baked in a vacuum oven at 100 ℃ for 8 h.

Preparation of IL: first, pyridine (40.0 g) was added into a 250 mL round-bottomed flask that was sealed with turned-mouth rubber stopper. Then nitrogen was pumped and filled into the flask three times repeatedly. At room temperature and under magnetic stirring, 75.4 g of butane bromide was added into the flask using micro-syringe, and then heated to 50 ℃ using water-bath apparatus and stirred for 12 h. The product was added to 200 mL of ethyl acetate, the process of oscillation and washing operation were carried out five times. And then this product was put into 200 mL of acetonitrile for recrystallization. Finally, the product was dried in a vacuum oven at 80 ℃ for 12 h.

Preparation of CMC/MWCNT composite: first, 1 mg of MWCNTs and 20 g of IL were added into a 100 mL round-bottomed flask under ultrasonication for 1 h at 80°C. Then 20 g of IL containing 5 wt.% CMC was added to the above solution, and the mixture solution was sonicated for 1h, followed by adding 20 mL of deionized water. The reaction solution was filtrated with 0.22 μm mixed fiber membranes. The resulting black product was dispersed in deionized water again after leaching and repeated five times. Finally, the obtained CMC/MWCNT composites were distributed into 5 mL of deionized water.

A 10 mg·mL^{-1} CMC/MWCNT water solution was pre-pared by ultrasonication for 10 min. The GC electrode was polished using 0.3 μm and 0.05 μm alumina slurry in sequence, rinsed with deionized water, and sonicated in acetone, ethanol and deionized water for 4 min each procedure. This solution was deposited on to the clean surface of GC electrode using a micro syringe, and finally dried with an infrared lamp to obtain a finished GC electrode coated with a CMC/MWCNT composite.

The surface morphology of CMC/MWCNT composite was investigated by SUPRATM 55 FESEM. HRTEM was conducted with JEOL JEM-3010. Electrochemical performances were measured using a CHI650C electrochemical workstation (CHI Instrument Company, Shanghai China) with a three-electrode system which composes of either a glass carbon electrode (GCE) or a GCE coated with a CMC/MWCNT composite as the working electrode, a saturated calomel electrode (SCE) as the reference electrode, and a Pt wire as the counter electrode.

If melamine is ingested it can interact with cyanuric acid to form melamine cyanurate, a crystalline composite that has been linked to fatal kidney stones. It was recently reported in China that children suffered from kidney stones after being fed with milk powders mixed melamine. Melamine was added to the mixtures to inexpensively increase the protein concentration of the milk powders.

Current methods to measure melamine concentrations include chromatography-mass spectrometry, fluorometry, spectro-photometry, potentiometry, capillary electrophoresis, electrokinetic chromatography, and colorimetric techniques. Although these methods could examine melamine under certain conditions, they have several disadvantages including a lack of sensitivity, low reproducibility, expensive apparatuses, complicated pretreatment, and time consuming procedures. It is necessary to develop a simple, specific, rapid and low-cost method for melamine detection in food and milk. Compared to conventional methods, electrochemical methods have become more important due to their simplicity, low-cost, accuracy, sensitivity and high stability. Cao et al. first reported an electrochemical method for determination of melamine using oligonucleotides modified gold electrodes by the interactions between melamine and oligonucleotides including electrostatic interaction and hydrogen-bonding. Cao et al. developed an electrochemical sensor for melamine detection based on that melamine may interact with 3,4-dihydroxyphenylacetic acid to form a complex mainly through the hydrogen bonding interaction. CNTs have unique electrical and mechanical property and large specific surface area, but the hydrophobicity of the CNTs limits their applications. Chitosan (CS) is an attractive biocompatible, biodegradable and non-toxic nature polysaccharide with abundant $-NH_2$ and $-OH$ functional groups, and can be used to prepare CNT/CS composite for widening the application of CNTs. As a new biocomposite, CNT/CS composite has excellent mechanical, biological, photoelectric properties because of its unique structure. CNT/CS composite modified electrode which has good responsibility, sensitivity, stability and reproducibility can be prepared easily and cheaply, and is widely applied in the fields of electrochemistry and biosensor.

In this section, we developed a simple and cheap electrochemical determination method for melamine using a glassy carbon (GC) electrode coated with a multi-wall CNT/CS composite. Furthermore, this method can be directly applied to the determination of melamine in real milk, and did not need many complicated pretreatments such as centrifugation and filtrate and the use of toxic solvent (e.g. trichloroacetic acid and methanol). Therefore, it is a simple and environmental protection process.

Melamine (analytical grade) was purchased from Tianjin Bodi Chemical Industry Co., Ltd. MWCNTs with 95% purity were supplied by Beijing Nano Technology. CS with a molecular weight of 5×10^4 and 90% deacetylation was purchased from Zhejiang Jinke Biochemical Co., Ltd. Nitric acid (analytical purity, 65%-68%) and sulfuric acid (analytical purity, 95%-98%) were purchased from Tianjin Third Chemical Factory. Acetic acid (analytical purity, 99.5%) was purchased from Xi'an Chemical Factory. Phosphate buffer solution (PBS, pH 6.9, 6.3×10^{-3} mol/L) was prepared for electrochemical measurements. Other chemicals were of analytical reagent grade and used as received. All aqueous solutions were prepared with deionized water.

Fourier transform infrared (FT-IR) spectrum was obtained using an EQUINOX-55 spectrum instrument. The surface morphology of MWCNT/CS composite was investigated by JEOL JSM-6700F FESEM. HRTEM was conducted with JEOL JEM-3010 at an accelerating voltage of 300 kV. Electrochemical measurements were carried out using a CHI 650D electrochemical workstation (CHI Instrument Company, Shanghai China) with a three-electrode system which was composed of either a GC electrode or a GC electrode coated with a MWCNT/CS composite as the working electrode, a saturated calomel electrode (SCE) as the reference electrode, and a Pt wire as the counter electrode.

At present, the normal synthesis methods of CNT/HAP composite are mechanical ball milling, ultrasonic dispersion, in situ method, chemical precipitation and hydrothermal treatment process. Santosh Aryal et al. reported the preparation of HAP/MWCNT composite by chemical deposition using CNTs-COOH. Zhao et al. reported the synthesization of HAP/CNT composite by self-assembly method using single wall CNTs as a template. Zhao et al. reported the preparation of HAP-CNTs powders by in situ deposition and hydrothermal treatment using dispersion agent. Sun et al. presented that HAP/CNT composite can be synthesized by ball milling, ultrasonication and in situ formation.

Although reported research in previous literatures was conducted with CNTs and HAP, the electrochemical performance of H_2O_2 with a glass carbon electrode coated with MWCNT/HAP composite has not been studied deeply. In this work, we synthesized biocompatible material HAP/MWCNTs, and utilized them to electrochemically detect H_2O_2 by using CV technique. The effect of MWCNT addition on the synthesis of HAP/MWCNT composite was optimized. The proposed research is expected to find the wide applications of HAP/MWCNT composite in biomedical area.

Carboxylic MWCNTs (COOH - MWCNTs) were prepared using a mixed solution ($HNO_3 : H_2SO_4 = 1 : 3$, V/V) by sonication for 4 h at 60 ℃. A $0.6 \text{ g} \cdot L^{-1}$ suspending solution was obtained with carboxylic MWCNTs and 1‰ sodium dodecylbenzene sulfonate (SDBS) under the condition of sonication for 2 h. The $0.4 - 3 \text{ mol} \cdot L^{-1} Ca^{2+}$ and $0.24 - 1.8 \text{ mol} \cdot L^{-1} PO_4^{3-}$ solutions were synthesized by using $Ca(NO_3)_2$ and $(NH_4)_2HPO_4$ respectively. $Ca(NO_3)_2$ solution was slowly added to the suspending solution at pH = 10. Under continuous stirring, a solution ratio of 3 : 5 between $(NH_4)_2HPO_4$ and $Ca(NO_3)_2$ dropped into a reactor chamber and the aging process was carried out for 24 h at 25 ℃. A HAP/MWCNT composite was obtained by in situ precipitation method after baking in the oven at 80 ℃ for 20 h.

$1 \text{ mg} \cdot mL^{-1}$ HAP/MWCNT ethanol solution was prepared by adding HAP/MWCNT composite to ethanol solution and being sonicated for 30 min. The GC electrode was polished using 0.3 μm and 0.05 μm alumina slurry in sequence and rinsed with deionized water, sonicated with acetone, ethanol and deionized water for 4 min each procedure. This solution was deposited on to the clean surface of GC electrode using a micro syringe, and finally dried with an infrared lamp to obtain a finished GC electrode coated with a HAP/MWCNT composite.

Activities and problems for students

Activities.

(1) How to functionalize CNTs?

(2) How to disperse CNTs?

Problems.

Disscuss the applications of carbon nanotubes in energy conversion.

Chapter 6　Graphene

6.1　History and discovery of graphene

The term "graphene" first appeared in 1987 in order to describe single sheets of graphite as one of the constituents of graphite intercalation compounds (GICs), conceptually a GIC is a crystalline salt of the intercalant and graphene. The term was also used in early descriptions of carbon nanotubes (CNTs), epitaxial graphene, and polycyclic aromatic hydrocarbons. However, none of these examples constitute isolated and two-dimensional graphene.

Larger graphene molecules or sheets (so that they can be considered as true isolated 2-D crystals) cannot be grown even in principle. An article in Physics Today reads: "Fundamental forces place seemingly insurmountable barriers in the way of creating (2-D crystals) ... Nascent 2-D crystallites try to minimize their surface energy and inevitably morph into one of the rich variety of stable 3-D structures that occur in soot. But there is a way around the problem. Interactions with 3-D structures stabilize 2-D crystals during growth. So one can make 2-D crystals sandwiched between or placed on top of the atomic planes of a bulk crystal. In that respect, graphene already exists within graphite ... One can then hope to fool nature and extract single-atom-thick crystallites at a low enough temperature that they remain in the quenched state prescribed by the original higher-temperature 3-D growth."

Single layers of graphite were previously (starting from the 1970s) grown epitaxially on top of other materials. This "epitaxial graphene" consists of a single-atom-thick hexagonal lattice of sp^2-bonded carbon atoms, as in free-standing graphene. However, there is significant charge transfer from the substrate to the epitaxial graphene, and in some cases, hybridization between the d orbits of the substrate atoms and π orbits of graphene, which significantly alters the electronic structure of the epitaxial graphene.

Single layers of graphite were also observed by transmission electron microscopy (TEM) within bulk materials, in particular inside soot obtained by chemical exfoliation. There have also been a number of efforts to make very thin films of graphite by mechanical exfoliation (starting from 1990 and continuing until after 2004) but nothing thinner than 50 to 100 layers was produced during these years.

The previous efforts did not result in graphene as we know it now, i.e. as "free standing" single-atom-thick crystals of a macroscopic size which are either suspended or interact only weakly with a substrate. It is not important whether graphene is suspended or placed on another (non-binding) substrate. In both cases, it is isolated and can be studied as such. Within this definition of graphene, it was first isolated by the Manchester group of Andre Geim who in 2004 finally managed to extract single-atom-thick crystallites from bulk graphite. He provided the first and unexpected proof for the existence of true (free-standing) 2-D crystals. Previously, it was assumed that graphene cannot exist in the flat state and should scroll into nanotubes "to decrease the surface energy".

This experimental discovery of 2-D crystal matter was openly doubted until 2005 when in the same issue of Nature the groups of Andre Geim and Philip Kim of Columbia University have proved "beyond a reasonable doubt" that the obtained graphitic layers exhibit the electronic properties prescribed by theory. This theory was first explored by Philip R Wallace in 1947 as a starting point for understanding the electronic properties of more complex, 3 - D graphite. The emergent massless Dirac equation was first pointed out by Gordon W. Semenoff and David P. DeVincenzo and Eugene J. Mele. Semenoff emphasized the occurrence in a magnetic field of an electronic Landau level precisely at the Dirac point. This level is responsible for the anomalous integer quantum Hall effect observed by the Manchester and Columbia groups. This was spelled out in detail by Valeriy P. Gusynin and Sergei G. Sharapov in a letter entitled "Unconventional Integer Quantum Hall Effect in Graphene".

Later, graphene crystals obtained by using theManchester recipe were also made suspended and their thickness proved directly by electron microscopy.

6.2 Mother of all graphitic forms

6.2.1 What's graphene?

Graphene is a one-atom-thick planar sheet of sp^2-bonded carbon atoms that are densely packed in a honeycomb crystal lattice.

It can be viewed as anatomic-scale "chicken wire" made of carbon atoms and their bonds. The name comes from "GRAPHITE" and "ENE", graphite itself consists of many graphene sheets stacked together.

The carbon-carbon bond length in graphene is approximately 0.142 nm. Graphene is the basic structural element of some carbon allotropes including graphite, CNTs and fullerenes (see Fig. 6 - 1). It can also be considered as an infinitely large aromatic molecule, the limiting case of the family of flat polycyclic aromatic hydrocarbons.

Chapter 6　Graphene

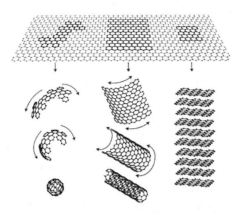

Fig. 6-1　As the mother of all graphitic forms, graphene is a 2-D building material for carbon materials of all other dimensionalities. Graphene can be wrapped up into 0-D buckyballs, rolled into 1-D nanotubes and stacked into 3-D graphite(from A. K. Geim and K. S. Novoselov. *Nature Materials*, 2007)

Graphene is a 2-D building block for all other dimensionalities. It can be wrapped up into 0-D buckyballs, rolled into 1-D nanotubes and stacked into 3-D graphite.

Graphene is the thinnest materials (about 0.335 nm) ever known in the world, and cannot be found until 2004 mainly because of the limitation of characterization methods. Currently only a few advanced technologies can be used to clearly characterize the morphology and structure of graphene, such as atomic force microscopy (AFM), scanning electron microscopy (SEM), TEM and Raman spectroscopy (see Fig. 6-2).

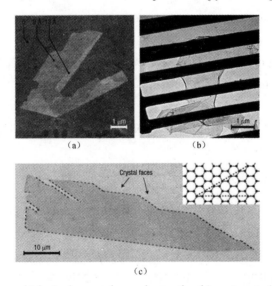

Fig. 6-2　One-atom-thick single crystals-graphene: the thinnest material you will ever see
(a) Graphene visualized by atomic force microscopy. The folded region exhibiting a relative height of about 4Å clearly indicates that it is a single layer. (b) A graphene sheet freely suspended on a micrometer-size metallic scaffold. (c) Scanning electron micrograph of a relatively large graphene crystal, which shows that most of the crystal's faces are zigzag and armchair edges as indicated by blue and red lines and illustrated in the inset (T.J. Booth, K.S.N, P. Blake and A.K.G. unpublished work).
1-D transport along zigzag edges and edge-related magnetism are expected to attract significant attention.
(from A. K. Geim and K. S. Novoselov.*Nature Materials*, 2007)

6.2.2 Description of graphene

The perfect graphene consist exclusively of hexagonal cells, pentagonal and heptagonal cells constitute defects. If an isolated pentagonal cell is present, then the plane warps into a cone shape, insertion of 12 pentagons would create a fullerene. Likewise, insertion of an isolated heptagon causes the sheet to become saddle-shaped. Controlled addition of pentagons and heptagons would allow a wide variety of complex shapes to be made, for instance carbon nanobuds. Single-walled carbon nanotubes (SWCNTs) may be considered to be graphene cylinders, some have a hemispherical graphene cap (that includes 6 pentagons) at each end.

The IUPAC compendium of technology states: "previously, descriptions such as graphite layers, carbon layers, or carbon sheets have been used for the term graphene ... it is not correct to use for a single layer a term which includes the term graphite, which would imply a three-dimensional structure. The term graphene should be used only when the reactions, structural relations or other properties of individual layers are discussed". In this regard, graphene has been referred to as an infinite alternant (only six-member carbon ring) polycyclic aromatic hydrocarbon. The largest molecule of this type consists of 222 atoms and is 10 benzene rings across.

Graphene was officially defined in the chemical literature in 1994 by the IUPAC as follows:

A single carbon layer of the graphitic structure can be considered as the final member of the series naphthalene, anthracene, coronene, etc. and the term "graphene" should therefore be used to designate the individual carbon layers in graphite intercalation compounds. (Note: use of the term "graphene layer" is also considered for the general terminology of carbons).

6.3　Structure and properties of graphene

6.3.1　Atomic structure

The atomic structure of isolated, single-layer graphene was studied by TEM on sheets of graphene suspended between bars of a metallic grid. Electron diffraction patterns showed the expected hexagonal lattice of graphene. Suspended graphene also showed "rippling" of the flat sheet, with amplitude of about 1 nm. These ripples may be intrinsic to graphene as a result of the instability of two-dimensional crystals, or may be extrinsic, originating from the ubiquitous dirt seen in all TEM images of graphene. Atomic resolution real-space images of isolated and single-layer graphene on silicon dioxide substrates were obtained by scanning tunneling microscopy (STM). Graphene processed using lithographic techniques is covered by photoresist residue, which must be cleaned to obtain atomic-resolution images. Such residue may be the "adsorbates" observed in TEM images, and may explain the rippling of suspended graphene. Rippling of graphene on the silicon dioxide surface was determined by conformation of graphene to the underlying silicon dioxide, and not an intrinsic effect.

Graphene sheets in solid form (e.g. density >1 g/cm^3) usually show evidence in diffraction for graphite's 0.34 nm (002) layering. This is true even of some single layer carbon nanostructures. However, unlayered graphene with only (hk0) rings has been found in the core of presolar graphite onions. TEM studies show faceting at defects in flat graphene sheets, and suggest a possible role in this unlayered-graphene for two-dimensional dendritic crystallization from a melt.

> **The structure differences between graphene, graphane and graphyne.**

Graphane, Graphene and Graphyne.

It's the same style, such as C_2H_6 (ethane), C_2H_4 (ethene), C_2H_2 (ethyne).

In 2007, it has been theoretically predicted that a related structure, called graphane, could exist in a stable form. Graphane consists of a single-layer structure with fully saturated (sp^3 hybridization) carbon atoms with C－H bonds in an alternating pattern (up and down with relation to the plane defined by the carbon atoms). Its two most stable conformations are the so-called chair-like (H atoms alternating on both sides of the plane) and boat-like (H atoms alternating in pairs)(see Fig. 6－3).

Fig. 6－3 Structure of graphane in the chair conformation. The carbon atoms are shown in gray and the hydrogen atoms in white. The figure shows the hexagonal network with carbon in the sp^3 hybridization
(from J. O. Sofo et al. Physical Review B, 2007)

A third member of these two-dimensional planar carbon structures, called graphyne, has also been predicted to exist but up to now only molecular fragments have been synthesized. New families of SWCNTs are proposed and their electronic structures are investigated. These nanotubes (called graphynes) result from the elongation of covalent interconnections of graphite-based nanotubes by the introduction of －yne groups(see Fig. 6－4).

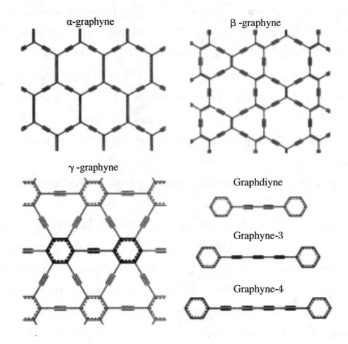

Fig. 6-4　Schematics of graphyne sheets used in MD simulations
(from Minmin Xue, et al. Nanotechnology, 2013)

6.3.2　Properties

1. Electronic properties

Graphene is quite different from most conventional three-dimensional materials. Intrinsic graphene is asemi-metal or zero-gap semiconductor. Understanding the electronic structure of graphene is the starting point for finding the band structure of graphite. It was realized early that the E-k relation is linear for low energies near the six corners of the two-dimensional hexagonal Brillouin zone, leading to zero effective mass for electrons and holes. Due to this linear "dispersion" relation at low energies, electrons and holes near these six points, two of which are inequivalent, behave like relativistic particles described by the Dirac equation for spin 1/2 particles. Hence, the electrons and holes are called Dirac fermions, and the six corners of the Brillouin zone are called the Dirac points. The equation describing the E-k relation is

$$E = h v_F \sqrt{k_x^2 + k_y^2}$$

where v_F, the Fermi velocity, is approximately 10^6 m/s.

The electrical properties of graphene can be described by a conventional Tight-Binding model, in this model the energy of the electrons with wavenumber k is

$$E = \pm \sqrt{\gamma^2 \left(1 + 4\cos^2 \frac{a \cdot k_y}{2} + 4\cos \frac{a \cdot k_y}{2} \cos \frac{\sqrt{3} a \cdot k_x}{2}\right)}$$

with the nearest-neighbor hopping energy $\gamma \approx 2.8$ eV and the lattice constant $a \approx 2.46$ Å. Conduction and valence band respectively correspond to the different signs in the above dispersion relation, they touch each other in six points, the "K-values". However, only two of these six points are independent, whereas the rest is equivalent by symmetry. In the vicinity of the K-points the energy depends linearly on the wavenumber, similar to a relativistic particle. Since an elementary cell of the lattice has a basis of two atoms, the wave function even has an effective 2-spinor structure. As a consequence, at low energies, even neglecting the true spin, the electrons can be described by an equation which is formally equivalent to the massless Dirac equation. Moreover, in the present case this pseudo-relativistic description is restricted to the chiral limit, i.e., to vanishing rest mass M_0, which leads to interesting additional features:

$$v_F \cdot \vec{\sigma} \cdot \nabla \psi(r) = E \cdot \Psi(r)$$

Here $v_F(10^6 \text{ m/s})$ is the Fermi velocity in graphene which replaces the velocity of light in the Dirac theory, $\vec{\sigma}$ is the vector of the Pauli matrices, $\Psi(r)$ is the two-component wave function of the electrons, and E is their energy.

2. Electronic transport property

The experimental results from transport measurements show that graphene has remarkably high electron mobility at room temperature, with reported values in excess of 15,000 cm$^2 \cdot V^{-1} \cdot $s^{-1}. Additionally, the symmetry of the experimentally measured conductance indicates that the mobility for holes and electrons should be nearly the same. The mobility is nearly independent of temperature between 10 K and 100 K, which implies that the dominant scattering mechanism is defect scattering. Scattering by the acoustic phonons of graphene places intrinsic limits on the room temperature mobility to 200,000 cm$^2 \cdot V^{-1} \cdot $s^{-1} at a carrier density of 10^{12} cm^{-2}. The corresponding resistivity of the graphene sheet would be 10^{-6} Ω·cm, less than the resistivity of silver, the lowest resistivity substance known at room temperature. However, for graphene on silicon dioxide substrates, scattering of electrons by optical phonons of the substrate is a larger effect at room temperature than scattering by graphene's own phonons, and limits the mobility to 40,000 cm$^2 \cdot V^{-1} \cdot $s^{-1}.

Despite the zero carrier density near the Dirac points, graphene exhibits a minimumconductivity in the order of $4e^2/h$ (where e is the elementary electric charge and h is Planck's constant). The origin of this minimum conductivity is still unclear. However, rippling of the graphene sheet or ionized impurities in the SiO_2 substrate may lead to local puddles of carriers that allow conduction. Several theories suggest that the minimum conductivity should be $4e^2/\pi h$; however, most measurements are of order $4e^2/h$ or greater and depend on impurity concentration.

Recent experiments have probed the influence of chemical dopants on the carrier mobility in graphene. Schedin et al. doped graphene with various gaseous species (some acceptors, some donors), and found the initial undoped state of a graphene structure can be recovered by gently heating the graphene in vacuum. Schedin et al. also reported that even for chemical dopant concentrations in excess of 10^{12} cm^{-2} there is no observable change in the carrier mobility. Chen et al. doped graphene with potassium in ultrahigh vacuum at low temperature. They found that potassium ions act as expected for charged impurities in graphene, and can reduce the mobility 20-fold. The mobility reduction is reversible by heating the graphene to remove the potassium.

3. Optical properties

Graphene's unique electronic properties produce an unexpectedly high opacity for an atomic monolayer, with a startlingly simple value: it absorbs $\pi\alpha \approx 2.3\%$ of whitelight, where α is the fine-structure constant. This has been confirmed experimentally (see Fig. 6-5), but the measurement is not precise enough to improve on other techniques for determining the fine-structure constant.

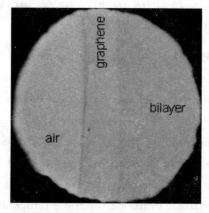

Fig. 6-5 Photograph of graphene in transmitted light. This one atom thick crystal can be seen with the naked eye because it absorbs approximately 2.3% of white light, which is π times fine-structure constant
(from Rahul Nair et al. *Science*, 2008)

Recently it has been demonstrated that the bandgap of graphene can be tuned from 0 – 0.25 eV (about 5 micron wavelength) by applying voltage to a dual-gate bilayer graphene field-effect transistor (FET) at room temperature. The optical response of graphene nanoribbons is tunable into the terahertz range by an applied magnetic field. Graphene/graphene oxide system exhibits electrochromic behavior, allowing tuning of both linear and ultrafast optical properties. A graphene-based Bragg grating (one-dimensional photonic crystal) has been fabricated and demonstrated its capability for excitation of surface electromagnetic waves in the periodic structure by using 633 nm He-Ne Laser as the light source.

4. Spin transport

Graphene is thought to be an ideal material for spintronics due to its small spin-orbit interaction and near absence of nuclear magnetic moments in carbon. Electrical spin-current injection and detection in graphene was recently demonstrated up to room temperature. Spin coherence length above 1 μm at room temperature was observed, and control of the spin current polarity with an electrical gate was observed at low temperature.

5. Anomalous quantum Hall effect

The quantum Hall effect is relevant for accurate measuring standards of electrical quantities, and Klaus von Klitzing received the Nobel Prize in 1985 for its discovery. The effect concerns the dependence of a transverse conductivity on a magnetic field, which is perpendicular to a current-carrying stripe. Usually the phenomenon, the quantization of the so-called Hall conductivity σ_{xy} at integer multiples of the basic quantity e^2/h can be observed only in very clean Si or GaAs solids, at very low temperatures around -270 ℃, and at very high magnetic fields.

Graphene in contrast, besides its high mobility and minimum conductivity, and because of certain pseudo-relativistic peculiarities to be mentioned below, shows particularly interesting behavior just in the presence of a magnetic field and just with respect to the conductivity-quantization, it displays an anomalous quantum Hall effect with the sequence of steps shifted by 1/2 with respect to the standard sequence, and with an additional factor of 4. Thus, in graphene the Hall conductivity is $\sigma_{xy} = \pm 4 \cdot (N+1/2)e^2/h$, where N is the above-mentioned integer "Landau level" index, and the double valley and double spin degeneracies give the factor of 4. Moreover, in graphene these remarkable anomalies can even be measured at room temperature, i.e. at roughly 20 ℃. This anomalous behavior is a direct result of the emergent massless Dirac electrons in graphene. In a magnetic field, their spectrum has a Landau level with energy precisely at the Dirac point. This level is a consequence of the Atiyah-Singer Index Theorem and is half-filled in neutral graphene, leading to the "$+1/2$" in the Hall conductivity. Bilayer graphene also shows the quantum Hall effect, but with the standard sequence (with $\sigma_{xy} = \pm 4 \cdot N \cdot e^2/h$), i.e. with only one of the two anomalies. Interestingly, concerning the second anomaly, the first plateau at $N=0$ is absent, indicating that bilayer graphene stays metallic at the neutrality point.

Unlike normal metals, the longitudinal resistance of graphene shows maxima rather than minima for integral values of the Landau filling factor in measurements of the Shubnikov-de Haas oscillations, which shows a phase shift of π, known as Berry's phase. The Berry's phase arises due to the zero effective carriers mass near the Dirac points. Study of the temperature dependence of the Shubnikov-de Haas oscillations in graphene reveals that the carriers have a non-zero cyclotron mass, despite their zero effective mass from the E-k relation.

6. Thermal properties

The thermal conductivity of graphene measured recently at the near-room temperature is between $(4.8\pm0.4)\times10^3$ to $(5.3\pm0.5)\times10^3$ W·m^{-1}·K^{-1}. These measurements, made by a non-contact optical technique, are in excess of those measured for carbon nanotubes or diamond. It can be shown by using the Wiedemann-Franz law, that the thermal conduction is phonon-dominated. However, for a gated graphene strip, an applied gate bias, causing a Fermi Energy shift much larger than $k_B T$ (k_B is Boltzman Constant, and T is Temperature), can cause the electronic contribution to increase and dominate over the phonon contribution at low temperatures.

Potential for this high conductivity can be seen by consideringgraphite, a 3-D version of graphene that has basal plane thermal conductivity of over a $1,000 \text{ W} \cdot \text{m}^{-1} \cdot \text{K}^{-1}$ (comparable to diamond). In graphite, the c-axis (out of plane) thermal conductivity is over a factor of 100 smaller due to the weak binding forces between basal planes as well as the larger lattice spacing. In addition, the ballistic thermal conductance of a graphene is shown to give the lower limit of the ballistic thermal conductance, per unit circumference, length of carbon nanotubes.

Despite its 2-D nature, graphene has 3 acoustic phonon modes. The two in-plane modes have a linear dispersion relation, whereas the out of plane mode has a quadratic dispersion relation. Due to this, the T^2 dependent thermal conductivity contribution of the linear modes is dominated at low temperatures by the $T^{1.5}$ contribution of the out of plane mode. The ballistic thermal conductance of graphene is isotropic.

7. Mechanical properties

As of 2009, graphene appears to be the strongest material ever tested. However, the process of separating it from graphite, where it occurs naturally, will require some technological development before it is economical enough to be used in industrial processes.

Utilizing an AFM, the spring constant of suspended graphene sheets have been measured. Graphene sheets, held together by van der Waals forces, were suspended over silicon dioxide cavities where an AFM tip was probed to test its mechanical properties. Its spring constant was in the range $1 - 5 \text{ N} \cdot \text{m}^{-1}$ and the Young's modulus was 0.5 TPa, which differs from that of the bulk graphite. These high values make graphene very strong and rigid. These intrinsic properties could lead to utilizing graphene for NEMS applications such as pressure sensors, and resonators.

As is true of all materials, regions of graphene are subject to thermal and quantum fluctuations in relative displacement. Although the amplitude of these fluctuations is bounded in 3-D structures, the Mermin-Wagner theorem shows that the amplitude of long-wavelength fluctuations will grow logarithmically with the scale of a 2-D structure, and would therefore be unbounded in structures of infinite size. Local deformation and elastic strain are negligibly affected by this long-range divergence in relative displacement. It is believed that a sufficiently large 2-D structure, in the absence of applied lateral tension, will bend and crumple to form a fluctuating 3-D structure. Researchers have observed ripples in suspended layers of graphene, and it has been proposed that the ripples are caused by thermal fluctuations in the material. As a consequence of these dynamic deformations, it is debatable whether graphene is truly a 2-D structure.

6.4 The synthesis methods of graphene

In the literatures, graphene has also been commonly referred to as monolayer graphite. This community has intensely studied epitaxial graphene on various surfaces (over 300 articles prior to 2004). In many cases, these graphene layers are coupled to the surfaces weak enough by van der Waals forces to retain the two dimensional electronic band structure of isolated graphene, as also happens with exfoliated graphene flakes with regard to silicon dioxide.

For example, experiments on epitaxial graphene monolayers onsilicon carbide, have provided the first demonstration of the spectrum of massless Dirac particles in graphene, which is the hallmark signature of its electronic structure. The weak van der Waals forces that provide the cohesion of multilayer graphene stacks do not always affect the electronic properties of the individual graphene layers in the stack. That is, while the electronic properties of certain multilayered epitaxial graphene are identical to that of a single graphene layer, in other cases the properties are affected as they are for graphene layers in bulk graphite. This effect is theoretically well understood and is related to the symmetry of the interlayer interactions.

In 2004, the physicists from University of Manchester found a way to isolate graphene by peeling it off from graphite with scotch tape and optically identify it by transferring them to a silicon dioxide layer on Si. It is now presumed that tiny fragments of graphene sheets are produced (along with quantities of other debris) whenever graphite is abraded, such as when drawing a line with a pencil.

Graphene produced by exfoliation is presently one of the most expensive materials onearth, with a sample, graphene can be placed at the cross section of a human hair costing more than \$1,000 as of April 2008 (about \$100,000,000 cm^{-2}). The price may fall dramatically, though, if commercial production methods are developed in the future. On the other hand, the price of epitaxial graphene on SiC is dominated by the silicon carbide substrate price which is approximately \$100 cm^{-2} as of 2009. This is about 1,000,000 times cheaper than exfoliated graphene.

1. Mechanical exfoliation method

The British researchers obtained relatively large graphene sheets (eventually, up to 0.1 mm in size and visible through a magnifying glass) by mechanical exfoliation (repeated peeling) of 3-D graphite crystals; their motivation was allegedly to study the electrical properties of thin graphite films and, as purely 2-D crystals were unknown before and presumed not to exist, their discovery of individual planes of graphite was presumably accidental. Both theory and experiment previously suggested that perfect 2-D structures could not exist in the free-state. It is believed that intrinsic microscopic roughening on the scale of 1 nm could be important for the stability of 2-D crystals.

Similar work is ongoing at many universities and the results obtained by the Manchester group in their PNAS paper "Two-dimensional atomic crystals" have been confirmed by several groups.

Without any chemical treatment, it is very simple to prepare high-quality graphene by mechanical exfoliation of graphite to study the structure and properties of graphene, and the discovery of graphene is also owe to this method. Of course, it is very hard and expensive to produce high-yield graphene with great size and definitely controlled layers.

2. Epitaxial growth on silicon carbide

Yet another method is to heatsilicon carbide at high temperature ($>1,100$ ℃) to reduce it to graphene. This process produces a sample size that is dependent upon the size of the SiC substrate used. The face of the silicon carbide used for graphene creation, the silicon-terminated or carbon-terminated, highly influences the thickness, mobility and carrier density of the graphene.

3. Epitaxial growth on ruthenium

This method uses the atomic structure of aruthenium substrate to seed the growth of the graphene (epitaxial growth). It doesn't typically yield a sample with a uniform thickness of graphene layers, and bonding between the bottom graphene layer and the substrate may affect the properties of the carbon layers.

4. Hydrazine reduction

Researchers have developed a method of placing graphene oxide paper in a solution of pure hydrazine (a chemical compound of nitrogen and hydrogen), which reduces the graphene oxide paper into single-layer graphene with higher electrical conductivity.

5. Sodium reduction of ethanol

A recent publication has described a process for producing gram-quantities of graphene, by the reduction of ethanol by sodium metal, followed bypyrolysis of the ethoxide product, and washing with water to remove sodium salts.

6. Chemical vapor deposition (CVD)

CVD method is usually used to prepare different thin films, the process to produce graphene is described as follows: transition metal substrates (such as nickel and copper) are deposited in the quartz tube full of hydrocarbon gas, then free carbon atoms are obtained because of the catalytic pyrolysis of hydrocarbon gas at high temperature, eventually the graphene at the surfaces of substrates are formed from free carbon atoms through cooling. A few factors vitally affect the size, layer, morphology and yield of the graphene, such as content of different gases, annealing temperature, reaction temperature, heating time, and cooling rate.

High-quality sheets of few-layer graphene exceeding 1 cm^2 in area have been synthesized via CVD on thin nickel layers. Moreover, large-area graphene films of the order of centimeters are prepared on copper substrates by CVD using methane. The films are predominantly single-layer graphene, with a small percentage (less than 5%) of the area having few layers, and are continuous across copper surface steps and grain boundaries (see Fig. 6-6). These sheets have been successfully transferred to various substrates, demonstrating viability for numerous electronic applications.

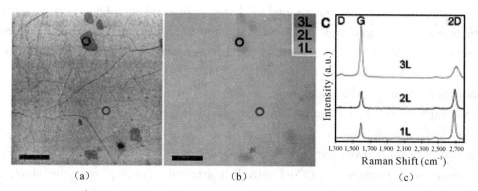

Fig. 6-6 The microstructure and Raman Spectra of graphene
(a) SEM image of graphene transferred on SiO_2/Si;
(b) Optical microscope image of the same regions as in (a);
(c) Raman spectra from the marked spots with corresponding colored circles or arrows showing the presence of one, two, and three layers of graphene

7. Arc discharge

The electric arc oven for synthesis of graphene mainly comprises two electrodes and a steel chamber cooled by water. The cathode and anode are both pure graphite rods. The current in the discharge process is maintained at 100-150 A. Up to now, the atmospheres for arc evaporation of graphite rods are H_2, NH_3 and He, air, etc. As the rods are brought close together, discharge occurs resulting in the formation of plasma. As the anode is consumed, the rods are kept at a constant distance from each other of about 1-2 mm by rotating the cathode. When the discharge ends, the soot generated is collected under ambient conditions. Only the soot deposited on the inner wall of the chamber is collected, avoiding the substance at the bottom of the chamber, for the latter tends to contain other graphitic particles. The arc-discharge method is useful to prepare pure, B- or N-doped graphene. B-doped graphene is obtained by carrying out the discharge in the presence of a mixture of a H_2 and B_2H_6. N-doped graphene is obtained by carrying out the discharge in the mixing atmosphere of He and NH_3 or H_2 and pyridine.

8. From CNTs

Experimental methods for the production of graphene ribbons are reported consisting of cutting open CNTs. In one such method MWCNTs are cut open in solution by action of potassium permanganate and sulfuric acid. In another method graphene nanoribbons are produced by plasma etching of nanotubes partly embedded in a polymer film.

6.5 Graphene growth with great size by CVD

We prepared graphene with great size (several centimeters) using CVD method by mainly controlling the reaction temperature and time, proposed a growth mechanism of graphene, and employed a simple method to characterize graphene. The effect of gas flow on the structure and morphology of graphene was studied. Moreover, in order to develop the applications of graphene in electrochemistry and biosensors, we studied the electrocatalytic activity of prepared graphene on the reduction of H_2O_2.

Copper foils with the thickness of 30 nm was pursued from Xi'an Dechi Electromechanical Co., Ltd. Argon (Ar, high purity), ethyne (C_2H_2, high purity), and hydrogen (H_2, high purity) were obtained from Shanghai Youjiali Co., Ltd. $Fe(NO_3)_3 9H_2O$ (analytical purity) was purchased from Tianjin Fuchen Chemical Factory. The graphene were prepared in OTF1200X-II quartz tube furnace made by Hefei Kejing Materials Technology Co., Ltd.

The smooth copper foils were put on ceramic plates in the quartz tube, and then were heated from room temperature to annealing temperature (800 ℃, 900 ℃ and 1,000 ℃). All samples underwent an annealing process at annealing temperature for 30 min with the flow of 500 sccm (standard-state cubic centimeter per minute) Ar and 220 sccm H_2, in order to increase the average Cu grain size and be good for the growth of graphene with large size. CVD growth of graphene was carried out under the flow of 500 sccm Ar, 220 sccm H_2, and 10 sccm C_2H_2, the reaction temperature varied from 800 ℃ to 1,000 ℃, and the reaction time varied from 1 min - 30 min. Afterwards, the temperature decreased rapidly from reaction temperature to 300 ℃ under gas flow of 500 sccm Ar and 220 sccm H_2 with the same cooling rate of 5 ℃ · min^{-1}, and then the temperature decreased slowly to the room temperature. Two pieces of graphene (3 cm × 2 cm) were added into 1 mL deionized water and then were sonicated for 10 s. to obtain the graphene solution. The glassy carbon electrode (GCE) was polished using 0.3 μm and 0.05 μm alumina slurry in sequence and rinsed with deionized water, and then was sonicated with acetone, ethanol and deionized water for 4 min each procedure. This graphene solution was deposited on the clean surface of GCE using a micro syringe, and finally dried with an infrared lamp to obtain a finished GCE coated with graphene.

The Haire DC-A 28 digital camera was used to obtain the optical digital photos. The optical microscope micrographs were obtained using a DMM-300C metallographic microscope (Shanghai Caikang Optical Instrument Co., Ltd). The surface morphology of graphene was investigated by FESEM, HRTEM, Raman spectroscopy (Horiba Jobin Yvon HR800). The Raman spectra were obtained using a He-Ne Laser with a wavelength of 514 nm and a laser power of 0.5 mW. Electrochemical measurements were carried out using a CHI 660D electrochemical workstation (CHI Instrument Company, Shanghai China) with a three-electrode system, which composed of either a GCE or a GCE coated with prepared graphene as the working electrode, a saturated calomel electrode (SCE) as the reference electrode, and a Pt wire as the counter electrode.

6.5.1 The transfer of graphene from copper foil to quartz slide

Fig. 6-7 shows the optical digital photos of graphene before and after the copper foil was totally dissolved, respectively. When graphene/copper foil was put into 0.5 M $Fe(NO_3)_3$ solution, copper foil was oxidized by $Fe(NO_3)_3$ and then dissolved, with the solution turned from yellow to faint yellow or even light green. The graphene (about 3.5 cm×2 cm) was absorbed on the quartz slide below the solution because of van der Waals force, but part of the graphene at the edge of the slide formed some wrinkles, which shows the graphene has good mechanical strength and flexibility. As shown in Fig. 6-7(b), four letters "NWPU" under the graphene were clearly seen, showing the good transparency of prepared graphene, which can be potentially applied to transparency electrodes.

Fig. 6-7 Optical digital photos of graphene before and after copper foil was totally dissolved respectively
(a)Before; (b)After

Fig. 6-8 shows the optical micrograph and SEM image of graphene respectively. As shown in Fig. 6-8(a) the graphene is also homogenous and the morphology of quartz slide under the graphene can be clearly seen, indicating the high transparency of graphene, which is consistent with Fig. 6-7. Meanwhile, we can see many wrinkles in the graphene. As shown in Fig. 6-8(b) the size of graphene is relatively big, and the carbon film is homogenous and swollen, and formed a number of wrinkles, which agrees with the optical digital photos.

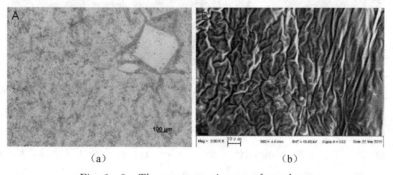

Fig. 6-8 The structure pictures of graphene
(a) optical micrograph; (b) SEM image of graphene
(reaction temperature: 800 ℃; reaction time: 30 min; C_2H_2: 10 sccm)

Table 6-1 shows the weight ratio (%) of chemical elements in graphene/copper and graphene. After copper being totally etched from the graphene, the carbon content is almost the same, the oxygen content increased greatly from 1.19% - 29.89%, the Fe content increased from 0 - 63.95%, and the copper content decreased from 94.87% to 3.09%. It can be explained that copper was totally oxidized, and generated $Fe(NO_3)_2$ and $Cu(NO_3)_2$ can be absorbed on the large surface of graphene.

Table 6-1 The weight ratio (%) of chemical elements in samples

Sample	C	O	Cu	Fe
graphene/copper	3.94	1.19	94.87	0
graphene	3.07	29.89	3.09	63.95

6.5.2 The effects of reaction time on graphene

Fig. 6-9 shows the optical digital photos of graphene prepared under different reaction times. Fig. 6-9 indicates that with the reaction temperature of 800 ℃ and C_2H_2 gas flow of 10 sccm, respectively, the size of graphene grew bigger gradually as reaction time increased from 1 min to 5 min. However when the reaction time increased from 5 min to 25 min, the size of graphene kept almost constant, but the graphene became darker and less transparent because of the gradually increased amorphous carbon particles and defects.

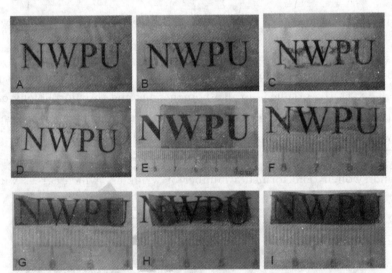

Fig. 6-9 Optical digital photos of graphene prepared by CVD at 800 ℃ (C_2H_2 gas flow of 10 sccm) under different reaction time (min) (A. 1, B. 2, C. 3, D. 4, E. 5, F. 10, G. 15, H. 20, I. 25)

Concerning the very low solubility of carbon in copper (0.01 wt.%, 1,084 ℃), many researchers verified the surface absorption mechanism of graphene grown on copper by the carbon isotope labeling, and believed that the nucleation and growth of the graphene domains on Cu substrate is determined by the supersaturation of active carbon derived from the decomposition of hydrocarbon by Cu catalyst. The lack of graphene domains can be attributed to the undersaturation of carbon precursors on the Cu surface. However, when the supersaturation reaches a certain critical point, graphene nuclei can form on the Cu catalyst and grow into graphene domains that can continuously grow until they merge together with the neighboring domains to fully cover the entire Cu surface.

In this case, at first only less active carbon form a few of graphene nuclei because of the decomposition of C_2H_2 by Cu catalysis. As the reaction time increased, the active carbon atoms in quartz tube increased and were absorbed more and more on the copper foils to reach supersaturation, and then much more graphene nuclei can form on the Cu catalyst and grow into graphene domains. So increasing reaction time resulted in more active carbon atoms and greater size of graphene until the Cu foils were fully covered after 5 min. That is to say, after 5 min the Cu surface were fully covered with graphene and the catalytic reaction of graphene was reduced severely, and the C_2H_2 cannot be fully catalyzed and decomposed, so the size of graphene kept constant with the reaction time exceeding 5 min. Our experimental results can also clearly explain the existence of surface absorption principle of graphene. Moreover, as the reaction time increased more than 5 min, due to the severe reduction of Cu catalysis, the more and more carbon from the incomplete decomposition of C_2H_2 formed more and more amorphous carbon particles and defects at the surface of copper foil, which made the graphene darker and less transparent. The similar phenomenon also existed when the reaction temperature was 900 ℃ and 1,000 ℃, respectively.

Fig. 6-10 shows the TEM images of graphene prepared under the reaction time of 15 min and 5 min, respectively. As shown in Fig. 6-10(a), graphene prepared with reaction time of 15 min can form some wrinkles because of van der Waals force and plenty of amorphous carbon particles were absorbed on the surface of graphene. Graphene was around 6 layers twice than that of the wrinkles (around 12 layers). Fig. 6-10(c) shows the structure diagram of graphene plane, and we can see that the distance of two nearest carbon atoms and of two nearest holes formed by six nearest carbon atoms in the graphene plane is around 0.14 nm and 0.24 nm, respectively. Fig. 6-10(b) is the TEM image of graphene prepared with reaction time of 5 min. In Fig. 6-10(b), carbon atoms formed honeycomb structure and each hole is surrounded by six carbon atoms. However, in Fig. 6-10(b) the crystalline structure of graphene is not whole and perfect, and graphene have some defects which may ascribe to long time bombardments of high energy transmission electrons. Because the distance of two nearest carbon atoms is very small, so we easily regard zigzag lines formed by adjacent carbon atoms as straight lines in TEM images. The distance of two neighboring lines in Fig. 6-10(b) is about 0.22 nm, which is consistent to theoretical distance of 0.21 nm in Fig. 6-10(c). Comparing Fig. 6-10(b) with Fig. 6-10(a), we can conclude that the graphene contained more defects (mainly amorphous carbon particles) as the reaction time increased, which agrees with Fig. 6-9.

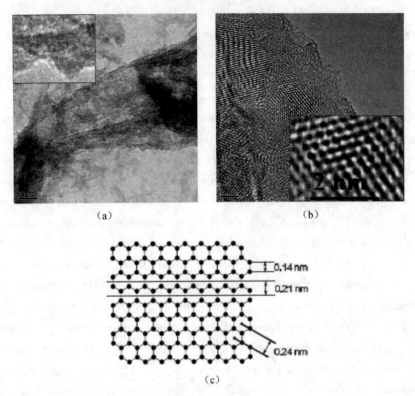

Fig. 6-10　TEM images of graphene (reaction temperature: 800℃) prepared with the reaction time of
(a) 15 min; (b) 5 min; (c) the structure diagram of graphene plane
(the inlets in (a) and (b) is the partial enlarged TEM images of (a) and (b), respectively)

6.5.3　The effects of reaction temperature on graphene

Fig. 6-11 is the optical photos of graphene prepared with different reaction temperatures. Fig. 6-11 display the carbon film got more transparent and amorphous carbon particles decreased as the reaction temperature increased from 800 ℃ to 1,000 ℃. The solid solubility of carbon atoms in copper is similar when the temperature is 800 ℃, 900 ℃ and 1,000 ℃ respectively, the amount of carbon atoms precipitated on the copper is almost the same. So the thickness of graphene grown from precipitated carbon atoms at different temperatures is almost the same. Moreover, at the same cooling rate of 5 ℃·min^{-1}, the cooling time will be longer at higher reaction temperature, and the amorphous carbon particles will hydrocrack and become fewer because of the existence of H_2. Besides, at higher temperature the absorbed amorphous carbon particles on the surface of copper foil will hydrocrack seriously. The similar phenomenon also happened when the reaction time was 10 min.

Fig. 6 - 11 Optical photos of graphene prepared with different reaction temperature
(reaction time: 5 min, C_2H_2 gas: 10 sccm)
(a)800 ℃; (b)900 ℃; (c)1,000 ℃

The G and D lines are assigned to the E_{2g} phonons of C sp^2 atoms and the breathing mode of κ-point phonons of A_{1g} symmetry activated by the presence of disorder, respectively. The most notable feature is the appearance of the 2-D peak, whose position and shape were shown to be related to the formation and the layer numbers of graphene. The intensity ratio of G- and 2-D-line (I_G/I_{2D}) is a sensitive probe of the number of graphene layers. The intensity ratio of D- and G-line (I_D/I_G) represents the degree of disorder of the graphite-based materials. Fig. 6 - 12 is Raman spectra of graphene prepared at different reaction temperatures. As shown in Fig. 6 - 12, I_D/I_G decreased and $I_G/I_{2D}(\approx 2.2)$ kept almost same value while the reaction temperature increased from 800 ℃ to 1,000 ℃, which indicates that with reaction temperature increasing the defects of graphene decreased, but the thickness of the graphene is almost the same, which agrees with the deduction from Fig. 6 - 11.

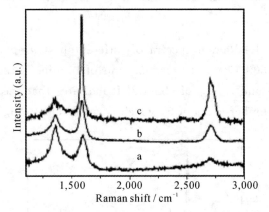

Fig. 6 - 12 Raman spectra of graphene prepared at different temperatures
(reaction time: 5 min, C_2H_2: 10 sccm, cooling rate: 5 ℃ · min^{-1})
a - 800 ℃; b - 900 ℃; c - 1,000 ℃

6.5.4 The effects of gas flow in the quartz tube on graphene

The effects of gas flow in the quartz tube on the graphene are also discussed. Fig. 6 - 13 shows the Raman spectra of different graphene prepared at the left and right areas of quartz tube, respectively. I_D/I_G of graphene in left area is higher than that of graphene in right area, but $I_G/I_{2D}(\approx 2.2)$ is almost the same, indicating the left graphene contains much more defects than right grapheme, and both of graphenes had the same layers. In the quartz tube, the gas flow moved from left to right, plenty of active carbon atoms were carried away from left to right, and the left tube was filled with major hydrocarbon, which can be absorbed on the surface of copper foil together with active carbon atoms, resulting to the generation of a great number of defects in the left graphenes. However, because the reaction temperature, time and cooling rate is the same, so the layers of the graphene prepared at different positions in the quartz tube is the same, which can be helpful for the production of graphene with the same layers in large scale.

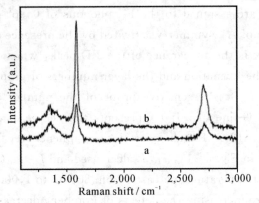

Fig. 6-13　Raman spectra of graphene prepared in different areas of quartz tube
(mixed gases flow from left to right in the quartz tube)
a – left;　b – right

Fig. 6-14 shows the Raman spectra of different part graphene prepared under the reaction temperature of 800 ℃ and the reaction time of 15 min. These two spectra are similar with I_D/I_G of about 0.3 and I_G/I_{2D} of about 3. It indicates that the graphene had the same defects, and the layer number is about 6, which is consistent to Fig. 6-10 (a).

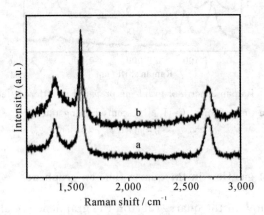

Fig. 6-14　Raman spectra of different part graphene (marked with "a" and "b", respectively)
(reaction temperature: 800 ℃, reaction time: 15 min)

6.5.5 The electrocatalytic properties of graphene

H_2O_2 are common raw materials and intermediate product in industrial procedure, and by-product of oxidase reaction in living organisms, so H_2O_2 was used to study the electrocatalytic properties of graphene in this paper. Fig. 6 - 15(a) shows the cyclic voltammograms (CVs) of (a) bare GCE and (b) GCE coated with graphene in 2.0×10^{-5} mol·L^{-1} phosphate buffer solution (PBS, 0.006 25 M, pH 6.9), respectively. The CV of bare GCE is almost a closed straight line, but the CV of GCE coated with graphene has an obvious reduction peak at around 0 V, because of the great specific surface area and good electrical conductivity of graphene, which cannot only increase the redox current but also decrease the redox over potential, effectively electrochemically catalyzing H_2O_2. Fig. 6 - 15 (b) shows the AC impedance curves of (a) bare GCE and (b) GCE coated with graphene in 2.0×10^{-5} mol·L^{-1} H_2O_2 PBS, respectively. The enrichment process of electrodes can be characterized by the change of impedance on the surface of electrodes. The typical Nyquist impedance curves contain two segments, that is to say, the semicircle segment at higher frequencies related to the electron transfer controlled process and linear segment at lower frequencies related to diffuse controlled process. In the impedance curves the semicircle diameter at higher frequencies is identified as electron transfer impedance and is related to the electron transfer kinetics of the redox probe on the electrode surface. As shown in Fig. 6 - 15(b) the electron transfer impedance of GCE coated with graphene is around 750 ohm, lower than that of bare GCE (1,500 ohm), and the impedance curves at lower frequencies is nearly a straight line. These results indicate electron transitivity and diffusivity of GCE coated with graphene is very good, due to the large specific surface area and good electrical conductivity of graphene, in agreement with Fig. 6 - 15(a).

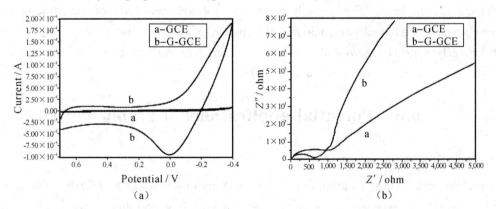

Fig. 6 - 15   The electrochemical performance of graphene
(a)CVs (scanning rate: 0.04 V·s^{-1}); (b) AC impedance curves of (a) bare GCE and (b) GCE coated with graphene in 2.0×10^{-5} mol·L^{-1} H_2O_2 PBS (pH 6.9), respectively

Fig. 6 - 16 shows CVs of GCE coated with graphene in 2.0×10^{-5} mol·L^{-1} H_2O_2 PBS (pH 6.9) at different scanning rates. With the increase of scanning rates the reduction peaks currents improved gradually and reduction peaks potentials shifted gradually towards left, which once again shows GCE coated with graphene has excellent electrocatalytic activity on the reduction of H_2O_2.

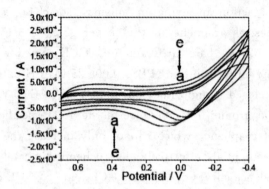

Fig. 6-16 CVs of GCE coated with graphene in 2.0×10^{-5} mol·L^{-1} H_2O_2 PBS(pH 6.9) at different scanning rates (V·s^{-1})

(a-0.02, b-0.04, c-0.06, d-0.12, e-0.20, respectively)

Chemical vapor deposition method was successfully used to prepare graphene several centimeters in size, promising for the production of graphene in large scale and potential application in the field of microelectronics. A very simple characterization method, optical digital camera was firstly carried out to successfully characterize the morphology of graphene, and optical micrograph XRD, SEM, TEM, and Raman spectra verified the validity of optical digital camera. Reaction temperature can seriously affect the defects of graphene but cannot affect the size and numbers of layers of graphene, and reaction time dominate the defects and size of graphene. Meanwhile, the gas flow in the quartz tube affects the defects of graphene. The growth mechanism of graphene and of amorphous carbon particles are also explained reasonably in this paper. Moreover, the electrocatalytic activity on the reduction of H_2O_2 was also studied, which will contribute to the applications of graphene in electrochemistry and biosensors.

6.6 Potential applications of graphene

1. Single molecule gas detection

Graphene can be used to prepare excellent sensors due to its 2-D structure. The fact that its entire volume is exposed to its surrounding makes it very efficient to detect adsorbed molecules. Molecule detection is indirect: as a gas molecule adsorbs to the surface of graphene, the location of adsorption experiences a local change in electrical resistance. While this effect occurs in other materials, graphene is superior due to its high electrical conductivity (even when few carriers are present) and low noise which makes this change in resistance readily detectable.

2. Graphene nanoribbons

Graphene nanoribbons (GNRs) are essentially single-layered graphene that are cut in a particular pattern resulting in certain electrical properties. Depending on how the un-bonded edges are configured, they can either be in a Z (zigzag) or Armchair configuration. Calculations based on tight binding predict that zigzag GNRs are always metallic while armchairs can be either metallic or semiconducting, depending on their width. However, recent density functional theory (DFT) calculations show that armchair nanoribbons are semiconducting with an energy gap scaling with the inverse of the GNR width. Indeed, experimental results show that the energy gaps do increase with decreasing GNR width. However, as of Feb. 2008, no experimental results have measured the energy gap of a GNR and identified the exact edge structure. Zigzag nanoribbons are also semiconducting and present spin polarized edges. Their 2-D structure, high electrical and thermal conductivity, and low noise also make GNRs a possible alternative to copper for integrated circuit interconnects. Some research is also being done to create quantum dots by changing the width of GNRs at select points along the ribbon, creating quantum confinement.

3. New graphene devices

Due to it sunique electronic property, graphene has also attracted the interest of technologists who see them as a way of constructing ballistic transistors. Graphene exhibits a pronounced response to perpendicular external electric fields allowing one to built FETs (field-effect transistors). In 2004, the Manchester group demonstrated FETs with a "rather modest" on-off ratio of 30 at room temperature. In 2006, Georgia Institute of Technology researchers announced that they had successfully built an all-graphene planar FET with side gates. Their devices showed changes of 2% at cryogenic temperatures. The first top-gated FET (on-off ratio of <2) was demonstrated by researchers of AMICA and RWTH Aachen University in 2007. Graphene nanoribbons may prove generally capable of replacing silicon as a semiconductor in modern technology.

Facing the fact that current graphene transistors show a very poor on-off ratio, researchers are trying to find ways for improvement. In 2008 researchers of AMICA and University of Manchester demonstrated a new switching effect in graphene field-effect devices. This switching effect is based on a reversible chemical modification of the graphene layer and gives an on-off ratio of greater than six orders of magnitude. These reversible switches could potentially be applied to nonvolatile memories.

In 2009 the Massachusetts Institute of Technology researchers built an experimental graphene chip known as a frequency multiplier. It is capable of taking an incoming electrical signal of a certain frequency and producing an output signal that is a multiple of that frequency. This graphene chip opens up a range of new applications, leading to various communications systems that can transmit data much faster than standard silicon chips.

4. Integrated circuits

Graphene has the ideal properties to be an excellent component of integrated circuits. Graphene has a high carrier mobility, as well as low noise allowing it to be utilized as the channel in a FET. The issue is that single sheets of graphene are hard to produce, and even harder to make on top of an appropriate substrate. Researchers are looking into methods of transferring single graphene sheets from their source of origin (mechanical exfoliation on SiO_2/Si or thermal graphitization of a SiC surface) onto a target substrate of interest. In 2008, the smallest transistor so far, one atom thick, 10 atoms wide was made of graphene. IBM announced in December 2008 that it has fabricated and characterized graphene transistors operating at GHz frequencies. In May 2009 a team from Stanford University, University of Florida and Lawrence Livermore National Laboratory announced that they have created an n-type transistor, which means that both n- and p-type transistors have now been created with grapheme.

5. Transparent conducting electrodes

High electrical conductivity and optical transparency make graphene a candidate for transparent conducting electrodes, required for such applications as touchscreens, liquid crystal displays, organic photovoltaic cells, and organic light-emitting diodes. In particular, mechanical strength and flexibility of graphene are advantageous, compared with indium tin oxide, which is brittle, and graphene films may be deposited from solution over large areas.

6. Ultracapacitors

Due to the incredibly high surface area to mass ratio of graphene, one potential application of graphene is in the conductive plates of ultracapacitors. It is believed that graphene could be used to produce ultracapacitors with a greater energy storage density than is currently available.

7. Graphene biodevices

Graphene's modifiable chemistry, large surface area, atomic-thickness and molecularly-gatable structure make antibody-functionalized graphene sheets excellent candidate for mammalian and microbial detection and diagnosis.

6.7 The future of graphene

On October 5, 2010, the Nobel Prize in Physics was awarded to Andre Geim and Konstantin Novoselov from the University of Manchester for their work on graphene.

It is most certain that we will see many efforts to develop various approaches to graphene electronics. Whichever approach prevails, there are two immediate challenges. First, despite the recent progress in epitaxial growth of graphene, high-quality wafers suitable for industrial applications still remain to be demonstrated. Second, individual features in graphene devices need to be controlled accurately enough to provide sufficient reproducibility in their properties. The latter is exactly the same challenge that the Si technology has been dealing with successfully. For the time being, to make proof-of-principle nanometer-size devices, one can use electrochemical etching of graphene by scanning-probe nanolithography. More recently, graphene samples prepared on nickel films, and on both the silicon face and carbon face of silicon carbide, have shown the anomalous quantum Hall effect directly in electrical measurements. Graphitic layers on the carbon face of silicon carbide show a clear Dirac spectrum in angle-resolved photoemission experiments, and the anomalous quantum Hall effect is observed in cyclotron resonance and tunneling experiments. Even though graphene on nickel and on silicon carbide have both existed in the laboratory for decades, it was graphene mechanically exfoliated on SiO_2 that provided the first proof of the Dirac fermion nature of electrons in graphene.

Despite the reigning optimism about graphene-based electronics, "graphenium" microprocessors are unlikely to appear for the next 20 years. In the meantime, many other graphene-based applications are likely to come of age. In this respect, clear parallels with nanotubes allow a highly educated guess of what to expect soon.

The most immediate application for graphene is probably its use in composite materials. Indeed, it has been demonstrated that a graphene powder of uncoagulated micrometer-size crystallites can be produced in a way scaleable to mass production. This allows conductive plastics at less than one volume percent filling, which in combination with low production costs makes graphene-based composite materials attractive for a variety of uses. However, it seems doubtful that such composites can match the mechanical strength of their nanotube counterparts because of much stronger entanglement in the latter case.

Another enticing possibility is the use of graphene powder in electric batteries that are already one of the main markets for graphite. An ultimately large surface-to-volume ratio and high conductivity provided by graphene powder can lead to improvements in the efficiency of batteries, taking over from the carbon nanofibers used in modern batteries. Carbon nanotubes have also been considered for this application, but graphene powder has an important advantage of being cheap to produce.

One of the most promising applications for nanotubes is field emitters, and although there have been no reports yet about such use of graphene, thin graphite flakes were used in plasma displays (commercial prototypes) long before graphene was isolated, and many patents were filed on this subject. It is likely that graphene powder can offer even more superior emitting properties.

CNTs have been reported to be an excellent material for solid-state gas sensors, but graphene offers clear advantages in this particular direction. Spin-valve and superconducting field-effect transistors are also obvious research targets, and recent reports describing a hysteretic magnetoresistance and substantial bipolar supercurrents prove graphene's major potential for these applications. An extremely weak spin-orbit coupling and the absence of hyperfine interaction in ^{12}C-graphene make it an excellent if not ideal material for making spin qubits. This guarantees graphene-based quantum computation to become an active research area. Finally, we cannot omit mentioning hydrogen storage, which has been an active but controversial subject for nanotubes. It has already been suggested that graphene is capable of absorbing a large amount of hydrogen, and experimental efforts in this direction are duly expected.

Activities and problems for students

Activities.

(1) Was graphene discovered in 2004?
(2) How to prepare graphene in large scale ?
(3) How many synthesis methods of graphene are there?

Problems.

What's the unique properties of graphene?

Reference

[1] Corbett J, Mckeown P A, Peggs G N, et al. Nanotechnology: international developments and emerging products[J]. CIRP Annals - Manufacturing Technology, 2000, 49(2):523-545.

[2] Invernizzi N, Foladori G, Maclurcan D, et al. Nanotechnology's controversial role for the South[J]. Science Technology and Society, 2008, 13 (1): 123-148.

[3] Freitas R A. Some Limits to global ecophagy by biovorous nanoreplicators, with public policy recommendations[J]. Hispanic American Historical Review, 2010, 90 (3):551-552.

[4] Brown G C, Hafner R P. A "top - down" approach to the determination of control coefficients in metabolic control theory[J]. European Journal of Biochemistry,1990, 188 (2), 321-325.

[5] Kovvuru S K, Mahita V N, Manjunatha B S, et al. Nanotechnology: the emerging science in dentistry[J]. Journal of Orofacial Research, 2012, 2 (1):33-36.

[6] Binnig G, Rohrer H. Scanning tunneling microscopy[J]. Surface Science, 1983, 126: 236-244.

[7] Binnig G, Quate C F, Gerber C, et al. Atomic Force Microscope[J]. Physical Review Letters, 1986, 56: 930.

[8] Yazawa Y, Yoshida H, Hattori T. The support effect on platinum catalyst under oxidizing atmosphere: improvement in the oxidation - resistance of platinum by the electrophilic property of support materials[J]. Applied Catalysis A: General, 2002, 237 (1): 139-148.

[9] Jacobson M, Cooper A R, Nagy J. Explosibility of metal powders[M]. Washington: Bureau of Mines, 1964.

[10] Yousaf A S, Ali S. Why nanoscience and nanotechnology? What is there for us? [J]. Journal of Faculty of Engineering & Technology, 2010, 15(1):286-293.

[11] Narayan R J, Kumta P N, Sfeir C, et al. Nanostructured ceramics in medical devices: applications and prospects[J]. JOM, 2004, 56 (10): 38-43.

[12] Caruso G, Raudino G, Caffo M, et al. Nanotechnology platforms in diagnosis and treatment of primary brain tumors [J]. Recent Patents on Nanotechnology, 2010, 4 (2):119-124.

[13] Nel A, Xia T, Meng H, et al. Nanomaterial toxicity testing in the 21st century: use of a predictive toxicological approach and high - throughput screening[J]. Accounts of Chemical Research, 2013, 46(3):607.

[14] Jepsen T, Jepsen M. Health IT: a roundtable discussion with healthcare information technology executives[J]. It Professional, 2008, 10(2):53-55.

[15] Michio I, Kang F Y. Carbon materials science and engineering—from fundamentals to applications [M].Beijing: Tsinghua University Press, 2006.

[16] Lu P J, Yao N, So J F, et al. The earliest use of corundum and diamond, in prehistoric China[J]. Archaeometry, 2005, 47(1): 1-12.

[17] Knite M, Teteris V, Kiploka A. The effect of plasticizing agent on strain - induced change of electric resistivity of carbon - polyisoprene nano - composites[J]. Materials Science & Engineering C, 2003, 23(23):787-790.

[18] Burger J. Consistency among methods of assessing concerns about the Los Alamos National Laboratory[J]. Journal of Toxicology & Environmental Health Part A, 2003, 66(2):199-210.

[19] Geim A K. Graphene: status and prospects[J]. Science, 2009, 324(5934): 1530-1534.

[20] Nasibulin A G, Pikhitsa P V, Jiang H, et al. A novel hybrid carbon material[J]. Nature Nanotechnology, 2007, 2(3): 156-161.

[21] Endo M, Kroto H W. Formation of carbon nanofibers [J]. Journal of Physical Chemistry, 1992, 96(17): 6941-6944.

[22] Frondel C, Marvin U B. Lonsdaleite a hexagonal polymorph of diamond[J]. Nature, 1967, 214(5088): 587-589.

[23] Yamada S, Sato H. Physics - some physical properties of glassy carbon[J]. Nature, 1962, 193(4812): 261.

[24] Gamaly E G, Rode A V, Luther - Davies B. Formation of diamond - like carbon films and carbon foam by ultrafast laser ablation[J]. Laser and Particle Beams, 2000, 18 (2): 245-254.

[25] Bailey M W, Bex P A. Industrial diamond - a brief history, a long future[J]. Finer Points, 1995, 7(4): 35-39.

[26] Scott R, Liddell H G. A Greek - English lexicon [M]. Oxford: Clarendon Press, 1996.

[27] Pisanty A. The electronic structure of graphite: a chemist's introduction to band theory[J]. Journal of Chemical Education, 1991, 68(10): 804.

[28] Lavrakas V. Textbook errors: guest column. XII: the lubricating properties of graphite[J]. Journal of Chemical Education, 1957, 34(5): 240.

[29] Iijima S. Helical microtubes of graphite carbon[J]. Nature, 1991, 354(6348):56-58.

[30] Iijima S. Direct observation of the tetrahedral bonding in graphitized carbon black by high resolution electron microscopy[J]. Journal of Crystal Growth, 1980, 50(3): 675-683.

[31] Markovic Z, Trajkovic V. Biomedical potential of the reactive oxygen species generation and quenching by fullerenes (C_{60})[J]. Biomaterials, 2008, 29(26): 3561-3573.

[32] Moussa F, Trivin F, Ceolin R, et al. Early effects of C_{60} administration in Swiss mice: a preliminary account for in vivo C_{60} toxicity [J]. Fullerene Science & Technology, 1996, 4(1): 21-29.

[33] Marks N A, Mckenzie D R, Pailthorpe B A, et al. Microscopic structure of

tetrahedral amorphous carbon[J]. Physical Review Letters, 1996, 76(5): 768-771.

[34] Marks N A, Mckenzie D R, Pailthorpe B A, et al. Ab initio simulations of tetrahedral amorphous carbon[J]. Physical Review B, 1996, 54(14): 9703-9714.

[35] Lewis J C, Redfern B, Cowlard F C. Vitreous carbon as a crucible material for semiconductors[J]. Solid-State Electronics, 1963, 6(3): 251-254.

[36] Rode A V, Hyde S T, Gamaly E G, et al.Structural analysis of a carbon foam formed by high pulse-rate laser ablation[J]. Applied Physics A, 1999, 69(1): S755-S758.

[37] Nasibulin A G, Anisimov A S, Pikhitsa P V, et al. Investigations of nanobud formation[J]. Chemical Physics Letters, 2007, 446(1-3): 109-114.

[38] Wu X J, Zeng X C. First-principles study of a carbon nanobud[J]. ACS Nano, 2008, 2(7): 1459-1465.

[39] Lee C, Wei X D, Kysar J W, et al. Measurement of the elastic properties and intrinsic strength of monolayer graphene[J]. Science, 2008, 321(5887): 385-388.

[40] Correa A A, Bonev S A, Galli G. Carbon under extreme conditions: phase boundaries and electronic properties from first-principles theory[J]. Proceedings of the National Academy of Sciences of the United States of America, 2006, 103(5): 1204-1208.

[41] Johnston R L, Hoffmann R. Superdense carbon, C_8: supercubane or analog of gamma-silicon[J]. Journal of the American Chemical Society, 1989, 111(3): 810-819.

[42] Openov L A, Elesin V F. Prismane C_8: a new form of carbon [J]. Journal of Experimental and Theoretical Physics Letters, 1998, 68(9): 726-731.

[43] Harris P J F. Fullerene-related structure of commercial glassy carbons [J]. Philosophical Magazine, 2004, 84(29): 3159-3167.

[44] Townsend S J, Lenosky T J, Muller D A, et al. Negatively curved graphitic sheet model of amorphous-carbon[J]. Physical Review Letters, 1992, 69(6): 921-924.

[45] Journet C, Maser W K, Bernier P, et al. Large-scale production of single-walled carbon nanotubes by the electric-arc technique[J]. Nature, 1997, 388(6644): 756-758.

[46] Thess A, Lee R, Nikolaev P, et al. Crystalline ropes of metallic carbon nanotubes[J]. Science, 1996, 273(5274): 483-487.

[47] Han S, Yun Y, Park K W, et al. Simple solid-phase synthesis of hollow graphitic nanoparticles and their application to direct methanol fuel cell electrodes [J]. Advanced Materials, 2003, 15(22): 1922-1925.

[48] Jang J, Ha H. Fabrication of carbon nanocapsules using PMMA/PDVB core/shell nanoparticles[J]. Chemistry of Materials, 2003, 15(11): 2109-2111.

[49] Jang J, Lim B. Selective fabrication of carbon nanocapsules and mesocellular foams by surface-modified colloidal silica templating[J]. Advanced Materials, 2002, 14(19): 1390-1393.

[50] Shimada T, Ohno Y, Suenaga K, et al. Tunable field-effect transistor device with metallofullerene nanopeapods[J]. Japanese Journal of Applied Physics Part 1: Regular Papers Brief Communications & Review Papers, 2005, 44(1A): 469-472.

[51] Halford B. The world according to rick[J]. Chemical & Engineering News, 2006, 84 (41): 13.

[52] Mraz S J. A new buckyball bounces into town[J]. Machine Design, 2005, 77(4): 70-73.

[53] Buseck P R, Tsipursky S J, Hettich R. Fullerenes from the Geological Environment [J]. Science, 1992, 257 (5067): 215-217.

[54] Wagner H, Farkas L. Synthesis of carbon "onions" in water[J]. Nature, 2001, 414 (6863): 506-507.

[55] Li Y, Huang Y, Du S, et al. Structures and stabilities of C_{60}-rings[J]. Chemical Physics Letters, 2001, 335 (5-6): 524.

[56] Gopakumar G, Nguyen M T, Ceulemans A. The boron buckyball has an unexpected T_h symmetry[J]. Chemical Physics Letters, 2008, 450 (4-6): 175.

[57] Krätschmer W, Lamb L D, Fostiropoulos K, et al. Solid C_{60}: a new form of carbon [J]. Nature, 1990, 347(6291):354-358.

[58] Haufler R E, Conceicao J, Chibante L P F, et al. Efficient production of C_{60} (Buckminsterfullerene), C60H36, and the solvated buckide ion[J]. Cheminform, 1991, 22(12):8634-8636.

[59] Krätschmer W, Fostiropoulos K, Huffman D R. The infrared and ultraviolet absorption spectra of laboratory-produced carbon dust: evidence for the presence of the C_{60} molecule[J]. Chemical Physics Letters, 1990, 170(s2-3):167-170.

[60] Taylor R, Hare J P, Abdul-Sada A K, et al. Isolation, separation, and characterization of the fullerenes C_{60} and C_{70}: the third form of carbon [J]. Cheminform, 2010, 22(14): 29.

[61] Hawkins J M, Meyer A, Lewis T A, et al. Crystal structure of osmylated C_{60}: confirmation of the soccer ball framework[J]. Science, 1991, 252(5003):312-313.

[62] Balch A L, Catalano V J, Lee J W, et al. (.eta.2 - C70)Ir(CO)Cl(PPh3)2: the synthesis and structure of an organometallic derivative of a higher fullerene[J]. Journal of the American Chemical Society, 1991, 113(23):8953-8955.

[63] Beavers C M, Zuo T, Duchamp J C, et al. $Tb_3N@C_{84}$: an improbable, egg-shaped endohedral fullerene that violates the isolated pentagon rule[J]. Journal of the American Chemical Society, 2006, 128 (35): 11352-11353.

[64] Becker L, Poreda R J, Hunt A G, et al. Impact event at the permian-triassic boundary: evidence from extraterrestrial noble gases in fullerenes[J]. Science, 2007, 291 (5508): 1530-1533.

[65] Beck M T, Mándi G. Solubility of C_{60}[J]. Fullerene Science & Technology, 1997, 5 (2):291-310.

[66] Bezmel'nitsyn V N, Eletskii A V, Okun' M V. Fullerenes in solutions[J]. Physics-Uspekhi, 2015, 168(168):308.

[67] Talyzin A V, Engström I. C_{70} in benzene, hexane, and toluene solutions[J]. Journal of Physical Chemistry B, 1998, 102 (34): 6477-6481.

[68] Haddon R C, Hebard A F, Rosseinsky M J, et al. Conducting films of C_{60} and C_{70} by alkali - metal doping[J]. Nature, 1991, 350 (6316): 320 - 322.

[69] Hebard A F, Rosseinsky M J, Haddon R C, et al. Superconductivity at 18 K in potassium - doped C_{60}[J]. Nature, 1991, 350 (6319): 600 - 601.

[70] Tanigaki K, Ebbesen T W, Saito S, et al. Superconductivity at 33 K in $Cs_x Rb_y C_{60}$ [J]. Nature, 1991, 352 (6332): 222 - 223.

[71] Iwasa Y, Takenobu T. Superconductivity, Mott Hubbard states, and molecular orbital order in intercalated fullerides[J]. Journal of Physics: Condensed Matter, 2003, 15 (13): R495.

[72] Capone M, Fabrizio M, Castellani C, et al. Strongly correlated superconductivity[J]. Science, 2002, 296 (5577): 2364 - 2366.

[73] Dresselhouse M S, Dresselhause G, Eklund P C. Science of fullerenes and carbon Nanotubes [M]. New York: Academic Press, 1995.

[74] Thompson D. Nanotechnology: basic science and emerging technologies[J]. Gold Bulletin, 2002, 35(4):135 - 136.

[75] Cveticanin J, Dustebek J, Veljkovic M. Endohedral fullerenes of different elements [J]. Journal of Optoelectronics & Advanced Materials, 2006, 8(5):1892.

[76] Saito R, Dresselhaus M S, Dresselhaus G. Physical properties of carbon nanotubes [M]. London: Imperial College Press, 1998.

[77] Monthioux M, Kuznetsov V. Who should be given the credit for the discovery of carbon nanotubes? [J]. Carbon, 2006, 44 (9): 1621.

[78] Oberlin A, Endo M, Koyama T. Filamentous growth of carbon through benzene decomposition[J]. Journal of Crystal Growth, 1976, 32 (3): 335 - 349.

[79] Dresselhaus M S, Endo M. Relation of carbon nanotubes to other carbon Materials [M]. Berlin: Springer, 2001.

[80] Abrahamson J, Wiles P G, Rhoades Brian L. Structure of carbon fibers found on carbon arc anodes[J]. Carbon, 1999, 37 (11): 1873.

[81] Iijima S. Helical microtubules of graphitic carbon[J]. Nature, 1991, 354 (6348): 56 - 58.

[82] Mintmire J W, Dunlap B I, White C T. Are fullerene tubules metallic? [J]. Physical Review Letters, 1992, 68 (5): 631 - 634.

[83] Bethune D S, Klang C H, De V, et al. Cobalt - catalyzed growth of carbon nanotubes with single - atomic - layer walls[J]. Nature, 1993, 363 (6430): 605 - 607.

[84] Krätschmer W, Lamb L D, Fostiropoulos K, et al. Solid C_{60}: a new form of carbon [J]. Nature, 1990, 347 (6291): 354 - 358.

[85] Kroto H W, Heath J R, O'Brien S C, et al. C_{60}: Buckminsterfullerene[J]. Nature, 1985, 318 (6042): 162 - 163.

[86] Bakshi S R, Lahiri D, Agarwal A. Carbon nanotube reinforced metal matrix composites — a review[J]. International Materials Reviews, 2010, 55(1):41 - 64.

[87] Zhbanov I, Sinitsyn N I, Torgashov G V. Nanoelectronic devices based on carbon nanotubes[J]. Radio Physics and Quantum Electronics, 2004, 47(5-6): 435-452.

[88] Goel R K, Panwar A. Carbon nanotubes: a new approach in material science[J]. International Journal of Latest Research in Science and Technology, 2012, 1(2): 155-158.

[89] Kumar M, Ando Y. Chemical vapor deposition of carbon nanotubes: a review on growth mechanism and mass production[J]. Cheminform, 2010, 41(22):3739.

[90] Ebbesen T W, Ajayan P M. Large-scale synthesis of carbon nanotubes[J]. Nature, 1992, 358 (6383): 220-222.

[91] Balaz P, Boldizarova E, Jelen S Ficeriová J, et al. Thiosulfate leaching of gold from a mechanically activated CuPbZn concentrate[J]. Hydrometallurgy, 2002, 67(1-3): 37-43.

[92] José Y M, Miki Y M, Rendón L, et al. Catalytic growth of carbon microtubules with fullerene structure[J]. Appl. Phys. Lett, 1993, 62 (6): 657.

[93] Ishigami N, Ago H, Imamoto K, et al. Crystal plane dependent growth of aligned single-walled carbon nanotubes on sapphire[J]. J. Am. Chem. Soc., 2008, 130 (30): 9918-9924.

[94] Randall L, Vander W, Thomas M, et al. Diffusion flame synthesis of single-walled carbon nanotubes[J]. Chemical Physics Letters, 2000, 323(3-4): 217-223.

[95] Singer J M. Carbon formation in very rich hydrocarbon-air flames. I. studies of chemical content, temperature, ionization and particulate matter[J]. Symposium on Combustion, 1958, 7(1): 559-569.

[96] Yuan L M, Saito K, Pan C X, et al. Nanotubes from methane flames[J]. Chemical physics letters, 2001, 340 (3-4): 237-241.

[97] Duan H M, McKinnon J T. Nanoclusters Produced in Flames[J]. Journal of Physical Chemistry, 1994, 98 (49): 12815-12818.

[98] Murr L E, Bang J J, Esquivel E V, et al. Carbon nanotubes, nanocrystal forms, and complex nanoparticle aggregates in common fuel-gas combustion sources and the ambient air[J]. Journal of Nanoparticle Research, 2004, 6 (2/3): 241-251.

[99] Sen S, Puri I K. Flame synthesis of carbon nanofibers and nanofibers composites containing encapsulated metal particles[J]. Nanotechnology, 2004, 15 (3): 264-268.

[100] Baughman R H, Zakhidov A A, de Heer W A. Carbon nanotubes-the route toward applications[J]. Science, 2002, 297(5582): 787-792.

[101] Yu M F, Lourie O, Dyer M J, et al. Strength and breaking mechanism of multiwalled carbon nanotubes under tensile load[J]. Science, 2000, 287(5453): 637-640.

[102] Reibold M, Paufler P, Levin A A, et al. Materials-carbon nanotubes in an ancient Damascus sabre[J]. Nature, 2006, 444(7117): 286-286.

[103] Zhang M, Fang S L, Zakhidov A A, et al. Strong, transparent, multifunctional carbon nanotube sheets[J]. Science, 2005, 309(5738): 1215-1219.

[104] Dalton A B, Collins S, Munoz E, et al. Super-tough carbon-nanotube fibres—these extraordinary composite fibres can be woven into electronic textiles[J]. Nature, 2003, 423(6941): 703-703.

[105] Postma H W C, Teepen T, Yao Z, et al. Carbon nanotube single-electron transistors at room temperature[J]. Science, 2001, 293(5527): 76-79.

[106] Collins P C, Arnold M S, Avouris P. Engineering carbon nanotubes and nanotube circuits using electrical breakdown[J]. Science, 2001, 292(5517): 706-709.

[107] Tseng Y C, Xuan P Q, Javey A, et al. Monolithic integration of carbon nanotube devices with silicon MOS technology[J]. Nano. Lett., 2004, 4(1): 123-127.

[108] Gabriel J C P. Dispersed growth of nanotubes on a substrate: WO, 2004040671A2[P]. 2004-04-12.

[109] Armitage P N. Flexible nanostructure electronic devices: US, 20050184641A1[P]. 2005-08-25.

[110] Guldi D M, Rahman G M A, Prato M, et al. Single wall carbon nanotubes as integrative building blocks for solar-energy conversion[J]. Angew Chem-Int Edit., 2005, 117(13): 2051-2054.

[111] Liu Z, Tabakman S, Welsher K, et al. Carbon nanotubes in biology and medicine: In vitro and in vivo detection, imaging and drug delivery[J]. Nano. Res., 2009, 2(2): 85-120.

[112] Simmons T J, Hashim D, Vajtai R, et al. Large area-aligned arrays from direct deposition of single-wall carbon nanotube inks[J]. J. Am. Chem. Soc., 2007, 129(33): 10088-10089.

[113] Dillon A C, Jones K M, Bekkedahl T A, et al. Storage of hydrogen in single-walled carbon nanotubes[J]. Nature, 1997, 386(6623): 377-379.

[114] Ye Y, Ahn C C, Witham C, et al. Hydrogen adsorption and cohesive energy of single-walled carbon nanotubes[J]. Appl. Phys. Lett., 1999, 74(16): 2307-2309.

[115] Chambers A, Park C, Baker R T K, et al. Hydrogen storage in graphite nanofibers[J]. J. Phys. Chem. B, 1998, 102(22): 4253-4256.

[116] Bachmatiuk A, Borowiak P E, Jedrzejewski R, et al. Study on hydrogen uptake of functionalized carbon nanotubes[J]. Phys. Status Solidi B-Basic Solid State Phys., 2006, 243(13): 3226-3229.

[117] Reddy A L M, Ramaprabhu S. Hydrogen adsorption properties of single-walled carbon nanotube-nanocrystalline platinum composites[J]. Int. J. Hydrogen Energy, 2008, 33(3): 1028-1034.

[118] Owen J R. Rechargeable lithium batteries[J]. Chem. Soc. Rev., 1997, 26(4): 259-267.

[119] Che G L, Lakshmi B B, Fisher E R, et al. Carbon nanotubule membranes for electrochemical energy storage and production[J]. Nature, 1998, 393(6683): 346-349.

[120] Kim C, Yang K S, Kojima M, et al. Fabrication of electrospinning-derived carbon nanofiber webs for the anode material of lithium-ion secondary batteries[J]. Adv.

Funct. Mater., 2006, 16(18): 2393-2397.

[121] Frackowiak E, Gautier S, Gaucher H, et al. Electrochemical storage of lithium in multi-walled carbon nanotubes[J]. Carbon, 1999, 37(1): 61-69.

[122] Gao B, Bower C, Lorentzen J D, et al. Enhanced saturation lithium composition in ball-milled single-walled carbon nanotubes[J]. Chem. Phys. Lett., 2000, 327(1-2): 69-75.

[123] Tarascon J M. Key challenges in future Li-battery research[J]. Philosophical Transactions of the Royal Society A: Mathematical, Phys. Eng. Sci., 2010, 368(1923): 3227-3241.

[124] Alvarez G, Montiel H, Pena J A, et al. Detection of the magnetic and electric transitions by electron paramagnetic resonance and low-field microwave absorption in the magnetoelectric $Pb(Fe_{0.5}Ta_{0.5})O_3$[J]. J. Alloys Compd., 2010, 508(2): 471-474.

[125] Ando Y, Zhao X, Shimoyama H, et al. Physical properties of multi-walled carbon nanotubes[J]. Int. J. Inorg. Mater., 1999, 1(1): 77-82.

[126] Tabatabaie F, Fathi M H, Saatchi A, et al. Effect of Mn-Co and Co-Ti substituted ions on doped strontium ferrites microwave absorption[J]. J. Alloys Compd., 2009, 474(1-2): 206-209.

[127] Meena R S, Bhattachrya S, Chatterjee R. Complex permittivity, permeability and microwave absorbing properties of $(Mn_{2-x}Zn_x)$ U-type hexaferrite[J]. J. Magn. Magn. Mater., 2010, 322(19): 2908-2914.

[128] Florance E T. Application of Kubo response functions to radiative transport[J]. J. Quant. Spectrosc. Radiat. Transfer, 1964, 4(5): 713-721.

[129] Kong I, Ahmad S H, Abdullah M H, et al. Magnetic and microwave absorbing properties of magnetite-thermoplastic natural rubber nanocomposites[J]. J. Magn. Magn. Mater., 2010, 322(21): 3401-3409.

[130] Nakamura T, Nagahata R, Suemitsu S, et al. In-situ measurement of microwave absorption properties at 2.45 GHz for the polycondensation of lactic acid[J]. Polymer, 2010, 51(2): 329-333.

[131] Chowdhury S R, Chen Y, Wang Y, et al. Microwave-induced rapid nanocomposite synthesis using dispersed single-wall carbon nanotubes as the nuclei[J]. J. Mater. Sci., 2009, 44(5): 1245-1250.

[132] Wadhawan A, Garrett D, Perez J M. Nanoparticle-assisted microwave absorption by single-wall carbon nanotubes[J]. Appl. Phys. Lett., 2003, 83(13): 2683-2685.

[133] Micheli D, Apollo C, Pastore R, et al. X-band microwave characterization of carbon-based nanocomposite material, absorption capability comparison and RAS design simulation[J]. Compos. Sci. Technol., 2010, 70(2): 400-409.

[134] Meena R S, Bhattachrya S, Chatterjee R. Complex permittivity, permeability and wide band microwave absorbing property of La^{3+} substituted U-type hexaferrite[J]. J. Magn. Magn. Mater., 2010, 322(14): 1923-1928.

[135] Hu N, Karube Y, Yan C, et al. Tunneling effect in a polymer/carbon nanotube nanocomposite strain sensor[J]. Acta Mater., 2008, 56(13): 2929-2936.

[136] Othman M B H, Ramli M R, Tyng L Y, et al. Dielectric constant and refractive index of poly (siloxane - imide) block copolymer[J]. Mater. Des., 2011, 32(6): 3173-3182.

[137] Michielssen Y, Sager J M, Ranjithan S, et al. Design of lightweight, broad - band microwave absorbers using genetic algorithms[J]. IEEE Trans. Microw. Theory Tech., 1993, 41(6-7): 1024-1031.

[138] Meena R S, Bhattachrya S, Chatterjee R. Development of "tuned microwave absorbers" using U - type hexaferrite[J]. Mater. Des., 2010, 31(7): 3220-3226.

[139] Meena R S, Bhattachrya S, Chatterjee R. Complex permittivity, permeability and microwave absorbing studies of ($Co_{2-x}Mn_x$) U - type hexaferrite for X - band (8.2 - 12.4 GHz) frequencies[J]. Mater. Sci. Eng. B - adv, 2010, 171(1-3): 133-138.

[140] Watts P C P, Hsu W K, Barnes A, et al. High permittivity from defective multiwalled carbon nanotubes in the X - band[J]. Adv. Mater., 2003, 15(7-8): 600-603.

[141] Kim H M, Kim K, Lee C Y, et al. Electrical conductivity and electromagnetic interference shielding of multiwalled carbon nanotube composites containing Fe catalyst[J]. Appl. Phys. Lett., 2004, 84(4): 589-591.

[142] Iijima S. Helical microtubules of graphitic carbon[J]. Nature, 1991, 354(6348): 56-58.

[143] Esawi A M K, Farag M M. Carbon nanotube reinforced composites: potential and current challenges[J]. Mater. Des., 2007, 28(9): 2394-2401.

[144] Kocharova N, Aaritalo T, Leiro J, et al. Aqueous dispersion, surface thiolation, and direct self - assembly of carbon nanotubes on gold[J]. Langmuir, 2007, 23(6): 3363-3371.

[145] Edgar K J, Buchanan C M, Debenham J S, et al. Advances in cellulose ester performance and application[J]. Prog. Polym. Sci., 2001, 26(9): 1605-1688.

[146] Yun S, Kim J. Multi - walled carbon nanotubes - cellulose paper for a chemical vapor sensor[J]. B, Sens. Actuators B - Chem., 2010, 150(1): 308-313.

[147] Fukushima T, Kosaka A, Ishimura Y, et al. Molecular ordering of organic molten salts triggered by single - walled carbon nanotubes[J]. Science, 2003, 300(5628): 2072-2074.

[148] Welton T. Room - temperature ionic liquids. Solvents for synthesis and catalysis[J]. Chemical Reviews, 1999, 99(8): 2071-2084.

[149] Swatloski R P, Spear S K, Holbrey J D, et al. Dissolution of cellose with ionic liquids[J]. Journal of the American Chemical Society, 2002, 124(18): 4974-4975.

[150] Nadagouda M N, Varma R S. Microwave - assisted synthesis of crosslinked poly (vinyl alcohol) nanocomposites comprising single - walled carbon nanotubes, multi - walled carbon nanotubes, and Buckminsterfullerene [J]. Macromolecular rapid

communications, 2007, 28(7): 842-847.

[151] Fukushima T, Aida T. Ionic liquids for soft functional materials with carbon nanotubes[J]. Chemistry - A European Journal, 2007, 13(18): 5048-5058.

[152] Marcilla R, Curri M L, Cozzoli P D, et al. Nano-obects on a round trip from water to organics in a polymeric ionic liquid vehicle[J]. Small, 2006, 2(4): 507-512.

[153] Venkatasami G, Sowa J R. A rapid, acetonitrile-free, HPLC method for determination of melamine in infant formula[J]. Analytica Chimica Acta, 2010, 665 (2): 227-230.

[154] Ibáñez M, Sancho J V, Hernández F. Determination of melamine in milk-based products and other food and beverage products by ion-pair liquid chromatography-tandem mass spectrometry[J]. Analytica Chimica Acta, 2009, 649(1): 91-97.

[155] Rima J, Abourida M, Xu T, et al. New spectrophotometric method for the quantitative determination of melamine using Mannich reaction[J]. Journal of Food Composition and Analysis, 2009, 22(7-8): 689-693.

[156] Tsai I L, Sun S W, Liao H W, et al. Rapid analysis of melamine in infant formula by sweeping - micellar electrokinetic chromatography [J]. Journal of Chromatography A, 2009, 1216(47): 8296-8303.

[157] Hsu Y F, Chen K T, Liu Y W, et al. Determination of melamine and related triazine by-products ammeline, ammelide, and cyanuric acid by micellar electrokinetic chromatography[J]. Analytica Chimica Acta, 2010, 673(2): 206-211.

[158] Treacy M M J, Ebbesen T W, Gibson J M. Exceptionally high Young's modulus observed for individual carbon nanotubes[J]. Nature, 1996, 381(6584): 678-680.

[159] Hone J, Whitney M, Piskoti C, et al. Thermal conductivity of single-walled carbon nanotubes[J]. Physical Review B, 1999, 59(4): R2514-R2516.

[160] Peniche C, Argüelles-Monal W, Peniche H, et al. Chitosan: an attractive biocompatible polymer for microencapsulation [J]. Macromolecular Bioscience, 2003, 3(10): 511-520.

[161] Madihally S V, Matthew H W T. Porous chitosan scaffolds for tissue engineering [J]. Biomaterials, 1999, 20(12): 1133-1142.

[162] Janegitz B C, Marcolino L H, Campana S P, et al. Anodic stripping voltammetric determination of copper(II) using a functionalized carbon nanotubes paste electrode modified with crosslinked chitosan[J]. Sensors and Actuators B: Chemical, 2009, 142(1): 260-266.

[163] Tsai Y C, Chen S Y, Lee C A. Amperometric cholesterol biosensors based on carbon nanotube-chitosan-platinum-cholesterol oxidase nanobiocomposite[J]. Sensors and Actuators B: Chemical, 2008, 135(1): 96-101.

[164] Saha A, Roy B, Garai A, et al. Two-component thermoreversible hydrogels of melamine and gallic acid[J]. Langmuir, 2009, 25(15): 8457-8461.

[165] Aryal S, Bahadur K C R, Dharmaraj N, et al. Synthesis and characterization of

hydroxyapatite using carbon nanotubes as a nano - matrix[J]. Scripta Materialia, 2006, 54(2): 131-135.

[166] Geim A K, Novoselov K S. The rise of graphene[J]. Nature Materials, 2007, 6(3):183.

[167] Simpson C D, Brand J D, Berresheim A J, et al. Synthesis of a giant 222 carbon graphite sheet[J]. Chemistry (Weinheim an der Bergstrasse, Germany), 2002, 8(6):1424.

[168] Boehm H P, Setton R, Stumpp E. Nomenclature and terminology of graphite intercalation compounds (IUPAC Recommendations 1994)[J]. Pure & Applied Chemistry, 1994, 66(9):1893-1901.

[169] Meyer J C, Geim A K, Katsnelson M I, et al. The structure of suspended graphene sheets[J]. Nature, 2007, 446(7131):60-63.

[170] Carlsson J M. Graphene: buckle or break[J]. Nature Materials, 2007, 6(11):801-802.

[171] Fasolino A, Los J H, Katsnelson M I. Intrinsic ripples in graphene[J]. Nature materials, 2007, 6(11): 858-861.

[172] Ishigami M, Chen J H, Cullen W G, et al. Atomic structure of graphene on SiO_2 [J]. Nano Letters, 2007, 7(6): 1643-1648.

[173] Kasuya D, Yudasaka M, Takahashi K, et al. Selective production of single - wall carbon nanohorn aggregates and their formation mechanism[J]. The Journal of Physical Chemistry B, 2002, 106(19): 4947-4951.

[174] Bernatowicz T J, Cowsik R, Gibbons P C, et al. Constraints on stellar grain formation from presolar graphite in the murchison meteorite[J]. Astrophysical Journal, 1996, 472(2):760-782.

[175] Wallace P R. The band theory of graphite[J]. Physical Review, 1947, 71(9):622-634.

[176] Neto A H C, Guinea F, Peres N M R, et al. The electronic properties of graphene [J]. Reviews of Modern Physics, 2009, 81(1): 109.

[177] Semenoff G W. Condensed - matter simulation of a three - dimensional anomaly[J]. Physical Review Letters, 1984, 53(26): 2449.

[178] Novoselov K S, Geim A K, Morozov S V, et al. Two - dimensional gas of massless dirac fermions in graphene[J]. Nature, 2005, 438(7065): 197-200.

[179] Schedin F, Geim A K, Morozov S V, et al. Detection of individual gas molecules adsorbed on graphene[J]. Nature materials, 2007, 6(9): 652-655.

[180] Zhang Y, Tang T T, Girit C, et al. Direct observation of a widely tunable bandgap in bilayer graphene[J]. Nature, 2009, 459(7248): 820-823.

[181] Tombros N, Jozsa C, Popinciuc M, et al. Electronic spin transport and spin precession in single graphene layers at room temperature[J]. Nature, 2007, 448(7153): 571-574.

[182] Balandin A A, Ghosh S, Bao W, et al. Superior thermal conductivity of single - layer graphene[J]. Nano Letters, 2008, 8(3): 902-907.

[183] Frank I W, Tanenbaum D M, Zande A M V D, et al. Mechanical properties of suspended graphene sheets[J]. Journal of Vacuum Science & Technology B Microelectronics & Nanometer Structures, 2007, 25(6):2558-2561.

[184] Bolmatov D, Mou C Y. Graphene-based modulation-doped superlattice structures [J]. Journal of Experimental and Theoretical Physics, 2011, 112(1): 102-107.

[185] Novoselov K S, Geim A K, Morozov S V, et al. Electric field effect in atomically thin carbon films[J]. Science, 2004, 306(5696): 666-669.

[186] Geim A K, Kim P. Carbon wonderland[J]. Scientific American, 2008, 298(4): 90-97.

[187] Subrahmanyam K S, Panchakarla L S, Govindaraj A, et al. Simple method of preparing graphene flakes by an arc-discharge method[J]. The Journal of Physical Chemistry C, 2009, 113(11): 4257-4259.

[188] Huh S H. Thermal reduction of graphene oxide[M]. NewYork: InTech, 2011.

[189] Kosynkin D V, Higginbotham A L, Sinitskii A, et al. Longitudinal unzipping of carbon nanotubes to form graphene nanoribbons[J]. Nature, 2009, 458(7240): 872-876.

[190] Schedin F, Geim A K, Morozov S V, et al. Detection of individual gas molecules adsorbed on graphene[J]. Nature Materials, 2007, 6(9): 652-655.

[191] Barone V, Hod O, Scuseria G E. Electronic structure and stability of semiconducting graphene nanoribbons[J]. Nano letters, 2006, 6(12): 2748-2754.

[192] Han M Y, Özyilmaz B, Zhang Y, et al. Energy band-gap engineering of graphene nanoribbons[J]. Physical Review Letters, 2007, 98(20): 206805.

[193] Lemme M C, Echtermeyer T J, Baus M, et al. A graphene field-effect device[J]. IEEE Electron Device Letters, 2007, 28(4): 282-284.

[194] Echtermeyer T J, Lemme M C, Baus M, et al. Nonvolatile switching in graphene field-effect devices[J]. IEEE Electron Device Letters, 2008, 29(8): 952-954.

[195] Wang X, Zhi L, Müllen K. Transparent, conductive graphene electrodes for dye-sensitized solar cells[J]. Nano Letters, 2008, 8(1): 323-327.

[196] Eda G, Fanchini G, Chhowalla M. Large-area ultrathin films of reduced graphene oxide as a transparent and flexible electronic material[J]. Nature Nanotechnology, 2008, 3(5): 270-274.

[197] Stoller M D, Park S, Zhu Y, et al. Graphene-based ultracapacitors[J]. Nano Letters, 2008, 8(10): 3498-3502.

[198] Mohanty N, Berry V. Graphene-based single-bacterium resolution biodevice and DNA transistor: interfacing graphene derivatives with nanoscale and microscale biocomponents[J]. Nano Letters, 2008, 8(12): 4469-4476.